Marble & Tile

Marble & Tile

The Selection & Care of Stone and Tile Surfaces

Frederick M. Hueston

NTC Publishing Company

Manufactured in the United States of America
Library of Congress Catalog Card Number 96-068655
ISBN: 0-9652577-0-3

Cover Photo: Stoneworks™ by American Olean Tile Company, 7834 C.F. Hawn Freeway, Dallas, Texas 75217

NTC Publishing Company
1413 Haven Drive
Orlando, FL 32803

Publications by Frederick M. Hueston

Stain Removal Guide for Stone

Stone Restoration & Maintenance- Problems & Solutions

How to Polish & Restore Marble Vanity & Furniture Tops- (Video)

Additional Publications offered by NTC Publishing Company

The Stone & Tile Report Newsletter

Marble- Understanding a Product of Nature by Ted Olson

Web page- http://www.webcreations.com/marble

HOW TO CONTACT THE AUTHOR

Frederick M. Hueston provides seminars, training and consulting services for the stone and tile trades. Requests for information about these services, as well as inquiries about availability for speeches and seminars, should be directed to the address below. Readers of this book are also encouraged to contact the author with comments and ideas for future editions.

Frederick M. Hueston
National Marble & Stone Consultants Inc.
1053 Portsmouth Lane
Winter Park, Florida 32792
E-mail: FHueston@aol.com

ABOUT THE AUTHOR

Frederick M. Hueston is a certified stone restoration specialist and Architectural Conservator. He is the founder and president of National Marble & Stone Consultants Inc. Of Winter Park, Florida. He is also the director of "The National Training Center for Stone & Masonry Trades" and publishes *The Stone & Tile Report Newsletter.*

Mr. Hueston has served as technical advisor, consultant and expert witness for numerous projects across the US, Canada, Europe and Asia for Government and private industry. He is a highly demanded speaker and prolific writer. Author of three books as well as numerous articles appearing in trade journals across the world.

Mr. Hueston is an active member of The Marble Institute of America, The Association for Preservation Technology and The National Trust for Historic Preservation.

Table of Contents

Marble & Tile

INTRODUCTION

Are you the proud owner of a new marble or ceramic tile floor or shower? Perhaps you have an old stone floor that has become worn over time. Or you're designing a new kitchen, but can't decide between marble or ceramic? Are granite countertops better than marble? Why, or why not?

Once you've made your choice of materials, then what? What should you use to clean the surface, and how often does it need cleaning? Can a vacuum cleaner be used, or should you dust-mop only? How do you keep it shiny? What will happen if you spill something on it? Do you have to seal it, and with what? Will it scratch? Does it need to be polished, and how often? Can it be sanded like a wood floor? If you're using stone or ceramic on shower walls, how do you remove soap scum? What do you use to clean the grout? Are you afraid to touch the surface for fear of ruining it?

These are just a few of the questions that concern most owners of stone and ceramic tile. The care of marble, stone and ceramic is not difficult; the real problem is finding good information. Few books on the subject exist, and those that do are hard to find. This book is designed to answer your marble, stone and ceramic tile questions, and to serve as a guide for the selection, maintenance and protection of these materials. If you carefully study and apply the information presented in the following pages, you will enjoy the lasting beauty and added value that stone and ceramic tile can bring to your home or office.

CHAPTER 1

Marble? Granite?Limestone? Ceramic? Porcelain or What?

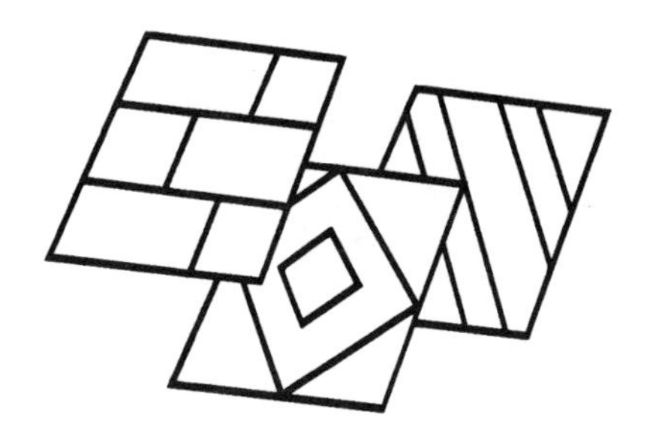

MARBLE? GRANITE? LIMESTONE? CERAMIC? PORCELAIN OR WHAT?

Sitting in my office one afternoon, I received a call from a frantic housewife.

"You've got to get over here right away," she said. "I spilled Kool-Aid on my marble floor and I think it's ruined!"

It was a slow afternoon, so I hopped in my truck and drove over to Mrs. Clark's house. When I arrived, she led me into the kitchen and pointed to a large red splatter-stain on the floor in front of the refrigerator.

"Is my marble ruined?" she asked, not sure she wanted to hear the answer.

I bent down, took a good look at it and replied, "No, your marble is not ruined. But you have one big mess here on your concrete tile floor."

"Concrete tile? I was told it was a marble floor!"

I carefully explained to her the difference between marble and concrete tile. The *good* news was, I managed to remove the stain, and she was happy after all.

This story represents just one example among hundreds of mistaken identity where being wrong may count—and cost a great deal. Perhaps the most important aspect of proper tile and stone selection and care is knowing the various types of tile and stone, and their characteristics and maintenance requirements . If you are making a decision about which type of tile to purchase, a through understanding of its care is an absolute must.

Why is this knowledge so important? Because certain cleaning chemicals that can be used very safely on some tile and stone can be extremely destructive on others. For example, there are several stone cleaners that are designed for cleaning granite; these chemicals typically contain acids, which are harmful to marble. If used on a polished marble surface, they will undoubtably dull it, and repairing the damage can be costly. So before you rush out and buy just any tile or stone cleaner, make sure you know exactly what type of material you're going to be dealing with. The following guide should help with identification; if you're still in doubt after reading it, consult your local stone and tile supplier.

And if you're still trying to decide which type of tile to install, please study the information in this book carefully. Knowing the characteristics and care of a material *before* you purchase it may prevent a catastrophic surprise in the future.

STONE & CERAMIC TYPES

MARBLE

There are over eight thousand types of marble on the market today, and the number continues to grow. It would be impossible to list every available type, but there are some common characteristics that make it fairly easy to identify marble.

Marble is commercially defined as *any limestone that will take a polish.* Limestones, and therefore marbles, are composed of minerals of calcite or dolomite. Marble in its purest state is white; colored marbles are the result of other minerals being mixed with the calcite or dolomite.

Now for those of you who don't really care what marble is made of, there are some other commonly recognizable characteristics:

- Marble, no matter what the color, will usually have some type of veining running through it; the veins are usually different in color then the main color of the stone. There are, however, exceptions to this rule. Some marbles, such as Thassos White will exhibit little or no veining.

Marble & Tile

• Marble is relatively soft when compared to other stones, such as granite. It will scratch very easily. If you run a knife blade lightly across the surface of the stone and it leaves a scratch, you are very likely to be dealing with marble. *Warning:* do not attempt this in the middle of your floor, if at all. If you need to determine how easily it will scratch, pick an inconspicuous spot, such as in a closet, under a carpet, etc. On dark marbles, these scratches will appear as light lines on the surface of the stone; on lighter-colored marbles it may be difficult to detect a scratch.

• Marble is also very sensitive to acidic chemicals. Vinegar, for example, is acidic, and will leave a dull spot on marble. It is extremely important to keep any marble surface out of contact with the following acidic materials: lemon, tomato and tomato sauce, bleach, coffee, fruit juices, wine, urine, vomit, tile cleaners like Tile X, the various mildew removers, X-14, acidic toilet bowl cleaners, cleaners containing lemon, pool pH decrease (muriatic acid), driveway cleaners, and, as noted, vinegar. Most products that contain acids will have their ingredients listed on the label. For easy reference, Table #1 lists the acids you may find on such products; avoid using them on or near any marble surface.

Foods and Chemicals to Avoid on Marble

The following foods and chemicals may contain acids and will etch a polished marble surface and leave a dull spot:

*VINEGAR * URINE *LEMON
*VOMIT *TOMATO * TILE X
*TOMATO SAUCE * MILDEW REMOVERS
*BLEACH * X-14 *COFFEE
*TOILET BOWL CLEANERS
*FRUIT JUICES
*LEMON CLEANERS *FRUITS
*POOL PH. DECREASE *WINE
*DRIVEWAY CLEANERS

Always check product labels for the presence of acids (see acid table).

Table #1

Acids to Avoid on Marble Surfaces

Any product containing the following acid should not be used on marble:

HYDROCHLORIC ACID	HCl
ACETIC ACID	CH_3COOH
SULFAMIC ACID	SO_3NH_3
PHOSPHORIC ACID	$H3PO_4$
SULFURIC ACID	H_2SO_4
OXALIC ACID	$H_2C_2O_4$
HYDROFLUORIC ACID	HF
NITRIC ACID	HNO_3
OTHER MISC ACIDS	

Table #2

GRANITE

Like marble, granite is a natural stone which occurs in many colors. There, however, the similarities stop. Unlike marble, granite is composed of different minerals with different properties. It is chiefly made up of 30% quartz and 60% feldspar; both substances are much harder than the calcite in marble, and for this reason, granite is much more difficult to scratch than marble. Granite is highly resistant to most acids, and will not etch and leave dull spots as they do on marble. There is only one acid to avoid: hydrofluoric acid, which is found in most rust removers. Granite rarely exhibits the veining characteristics of marble. It contains crystals which are very distinct, often giving the appearance of small to medium-sized stones compacted together . There are some exceptions. Blue Azul, for example, has a vein-like pattern, but if the veins are examined closely, they will also be seen to contain small, distinct crystals. Granite is an excellent choice for countertops, especially in the kitchen, where it will be exposed to acidic foods and chemicals.

LIMESTONE

Now, pay close attention. I'll try to make this as clear as possible. *Limestone is not a marble, but marble is a limestone.* Let me clarify this. *Limestones* are made up of calcite from shells, coral and other debris. They are what we call sedimentary rocks—that is, they have been formed by the breakdown of other rocks, shells, etc. *Marbles* are limestones that have been subjected to great heat and pressure and as a result, have changed (*metamorphosed*) into marble.

What difference does this make? There are several differences, and they are very important. Limestones are usually loosely held together, and may differ in porosity from marble. The coarse grain of some limestones give them excellent wearability. Limestones can contain numerous fossil impressions that are well preserved. If a piece of stone has various shell or animal-like patterns there's a very good chance that it's limestone. Limestones come in various colors, but most are shades of brown or tan, with some leaning toward gray and red. All seem to fall into the earth-tone color range. Limestone is becoming increasingly popular in the American West and Southwest.

SLATE

Slate is commonly gray in color, although you may find green, yellow and red hues. It is a stone which has been metamorphosed from shale—that is, it consists of clay-like materials that have undergone change under heat and pressure. For the layman, slate can be recognized by its sheet-like structure. The material is usually thin and if broken in half, will flake off into sheets. Slate is seldom highly reflective unless a coating is placed on it, and its surface is usually uneven unless machine- sanded.

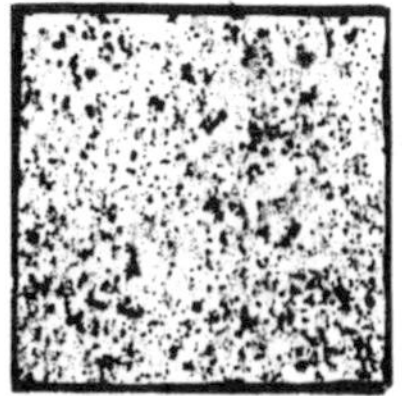

QUARTZITE

Quartzite is a rare material used for flooring which, like limestone, is gaining in popularity. It can be found in colors from white to a pinkish-brown. It is composed of metamorphosed quartz sand or sandstone. It is acid-resistant. Its texture appears sugary. It is commonly used on floors, patios, driveways etc.

SANDSTONE

Sandstone is seldom seen as a flooring material. It commonly is used for building stone, but occasionally makes its way indoors as a flooring surface. Composed primary of quartz, it is loose and rough in texture. As its name implies, it has the appearance of sand crystals cemented together. It is also acid-resistant, and very rarely is polished.

AGGLOMERATES

Agglomerate stones are composed of broken pieces of marble and sometimes granite bound together with a polyester or epoxy bonding material. Their properties are similar to those of the natural stones they are made of, but their plastic binders may be less shiny than the stone chips, giving the material an alligator skin-like appearance. Agglomerates are both very popular and inexpensive compared to purely natural stones, and their refinishing requirements are slightly different due to the plastic binders .

TERRAZZO

Terrazzo flooring was very popular in the late '50s and early '60s, and is making a big comeback in the 1990s. Terrazzo is a poured flooring material which is essentially Portland cement with the addition of marble chips. The marble-chip-and-cement mixture is poured and spread out, and the desired surface texture achieved by grinding it flat and applying a wax coating. The maintenance requirements for terrazzo are identical to those of marble. Terrazzo is more uniform in appearance than most other stones because it can be poured into very large areas without grout lines.

RIVER ROCK

River rock flooring , sometimes called Chattahoochee River Rock, is commonly used outdoors on patios and pool decks, but occasionally sees service as an indoor floor. River rock flooring consists of a layer of river rocks heavily coated with urethane or epoxy. This coating serves to hold the rocks in place and to provide a smooth surface.

FLAGSTONE

Flagstone is a term given to almost any stone material cut into thin, irregular shapes. It is usually sandstone, quartz, bluestone or slate, and most often is used as paving for walks, driveways and patios. It can sometimes be found indoors in foyers.

BLUESTONE

Bluestone gets its name from it bluish-gray hue. Found generally in the form of flagstone, it is a dense, fine-grained sandstone, which, like other flagstones, is typically used outdoors as a paver for walks and patios.

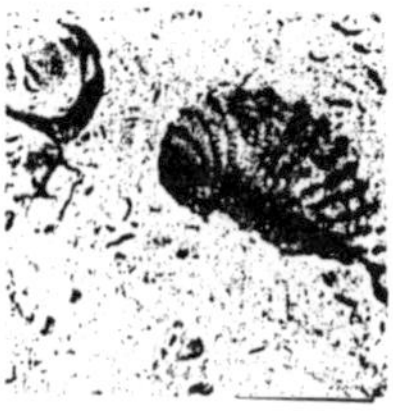

COQUINA
OR SHELL STONE

Coquina and/or shell stone is a limestone composed of broken fragments of shells and corals. This sedimentary rock, which is sometimes described as "coral stone," is extremely porous, wears relatively well and is rarely polished.

ONYX

Onyx is a type of marble which has been deposited from cold solutions. Characteristically translucent, with many veins running concentrically to one another. It is very expensive and is usually fashioned into small table tops. Its properties are similar to those of marble.

SOAPSTONE

Soapstone is a very soft mineral of talc. Rarely used for flooring, it is often fabricated into fireplace hearths and table tops, and is very popular as a medium for statuary.

TRAVERTINE

Travertine is a type of limestone, but differs from other forms in that it is formed in hot springs called *karst.* The water movement in these karst erodes the travertine, creating holes in the stone. Polished travertine will usually have its holes filled in with colored Portland cement. Since the fillers typically do not take a polish, they tend to give the stone a blotchy appearance. Unfilled travertine, which does exhibit these holes, is widely used on walls, floors and building exteriors. Travertine is commonly tan or beige in color, but can also be found in silver and reds.

CONCRETE TILE

Concrete tile is available in various sizes and an endless variety of colors. It can be painted to simulate natural stone and is often mistaken for the real thing. It can be an excellent substitute for natural stone for those on a tighter budget. Care must be taken in selection, since poorer-grade concrete tiles will tend to wear and discolor faster and more easily than most natural stones.

CULTURED MARBLE

Cultured marble is a man-made product composed of plastic resins. It is primarily used for countertops, but occasionally you may encounter cultured marble floor tiles. There are some excellent cultured marbles, and some are difficult to tell from the real thing. They are generally softer than natural stone and will scratch easily.

CORIAN™

Corain™ is manufactured by Dupont and is also a man-made plastic resin. It differs from cultured marble in that its surface can be repaired by sanding. It is a durable material used primarily for kitchen and bath countertops.

MEXICAN TILE/
TERRA COTTA

Mexican tile is a terra cotta-type material. Terra cotta is made from clay, baked in the sun or oven dried. Its color range is limited to brown, yellow and reds. Some terra cotta can be stained or pickled with dyes to change its color, but these dyes wear easily. Although Mexican tiles are typically unglazed, there are glazed tiles available.

CERAMIC TILE

Ceramic tile is made from a mixture of clays baked at very high temperatures. From tiny 1″ mosaics to 24″ x 24″ floor tiles, the variety of sizes and colors in which this material can be found is virtually limitless.

Ceramic tiles are available in glazed or unglazed finishes. A *glazed* finish is a hard surface—usually glossy or satin-like, but it can also be matte, semi-matte or dull—which is achieved during the firing process by applying a special material called a bisque and then baking the tile a second time. The color and patterns in most glazed ceramic tile contained is in this glazed layer. When you chip a ceramic tile, it is generally the glaze that you remove.

Unglazed ceramic tile does not contain this extra layer. Its finish is usually dull, and the tile is essentially the color of the clay used to make it. Unglazed ceramic tiles that do display a sheen or gloss have had a sealer or wax treatment applied to them.

QUARRY TILE

Quarry tile is a variety of unglazed ceramic tile. Commonly used in commercial kitchens and restaurants and produced in earth-tone colors, its rough texture makes it an excellent option for areas where slip resistance is a major concern.

PAVERS

Although technically a term which is interchangeable with "tile," *paver* usually refers to floor tile only. Pavers are similar to quarry tile, and are often rectangular rather than square.

PORCELAIN TILE

Porcelain tile is becoming increasingly popular as a durable floor tile. Like other types of ceramic tile, porcelain is baked. Unlike glazed ceramic, however, it does not have a bisque surface; the material, color included, is consistent all the way through the tile body. An extremely hard material, porcelain tile is almost impossible to chip, and is nearly indestructible. It is produced in formats as large as two feet square, and comes in a honed or polished finish. Light-colored porcelain can trap dirt and will need regular maintenance.

BRICK

Brick flooring is commonly found in commercial floors which are exposed to heavy traffic, but it can also lend a rustic appearance to a residential setting. Brick is available in several earth-tone colors, and comes in the form of full bricks or splits (half-thicknesses). Since brick flooring is difficult to keep clean, a topical sealer is highly recommended.

GROUT

Although grout is not a floor tile or paver, it is an important component of the floor. Grout is usually a cement-based material that is used to fill in the spaces between the tile. It is produced in a variety of colors.

There are basically two types of cementitious grout: sanded and unsanded. There are several latex additives that can be mixed with the grout during installation to provide stain resistance. It is advisable to seal your grout with a silicone sealer to prevent staining.

In addition to cement-based grouts there are epoxy grouts, which are made from plastic resins of epoxy and mixed with the grout at installation. They are usually more expensive than cementitious grouts, but are extremely stain resistant. Epoxy grouts are also available in many colors. The short guide which follows should assist you in selecting the proper grouting for your tile.

GROUT TYPES

Several options are open to you when choosing grout for your tile floor, wall or countertop, and a thorough understanding of them now may spare you some headaches later. Some of the worst problems associated with tile have nothing to do with the tile and *everything* to do with the grout. A beautiful white marble floor, for example, is hardly seen in its best light when the white grout is black and dirty. The basic types of grout are:

Sanded grout. This is the type most commonly used for ceramic tile, stone and any tile with a grout joint ⅛″ or larger. Composed of Portland cement, sand and other additives, it is mixed with water and troweled into the grout joint, where it takes approximately 24 hours to dry. Although it becomes as hard as concrete when fully cured, it can pose several problems. It is very absorbent, and if it is not sealed it will soak up stains, dirt and any other liquid spilled on the tile. Care should be exercised in choosing a good sealer to prevent staining and water absorption. Although many grouts can be mixed at the time of installation with a latex additive which will reduce absorbency, I would still recommend sealing. Another problem arises when it is used with marble. Since the grout is made with sand, it *will* scratch polished marble when installed. This is a common problem among tile contractors who are not familiar with stone installation. *Never use sanded grout on polished marble!* Polished marble should be installed with a grout width smaller than ⅛″.

Unsanded or wall grout. Unsanded grout, commonly called "wall grout," is essentially sanded grout without the sand. It is used on ceramic tile and polished marble with grout joints smaller than ⅛". All the cleaning problems associated with sanded grout apply to wall grout, which should be sealed after installation to reduce absorbency.

Latex-modified grout additives. Several latex additives are available that can be added to both sanded and unsanded grouts. These additives are blends of acrylics and latex, and will decrease water absorption, increase strength and improve color retention. Some grouts have dried latex powder added to them at the factory and do not require additional additives. A number of manufacturers also include anti-fungal and mildew-resistance additives.

Epoxy grout. This type of grout is a waterless two-part formula consisting of epoxy resins (Part A) and a hardener (Part B). These components are mixed on-site just prior to grouting. When fully cured, epoxy grouts are stain- and mildew-resistant; they are also less absorbent than cement-based grouts, and are easily cleaned. They should be used on tile and stone on kitchen countertops, backsplashes and bathrooms. Epoxy grouts are difficult to apply, and can be quite messy during application; be sure to hire a contractor who is skilled in their use. These grouts require no additional sealer.

Furan grout. Furan grouts are similar to epoxies, but are composed of polymers of furfuryl alcohol, which are highly resistant to chemical action. They are rarely used for residential installation, but are often employed in industrial projects, such as laboratories, dairies, meat-packing plants, etc. Furan grouts are only available in black, and special skills are required for proper installation.

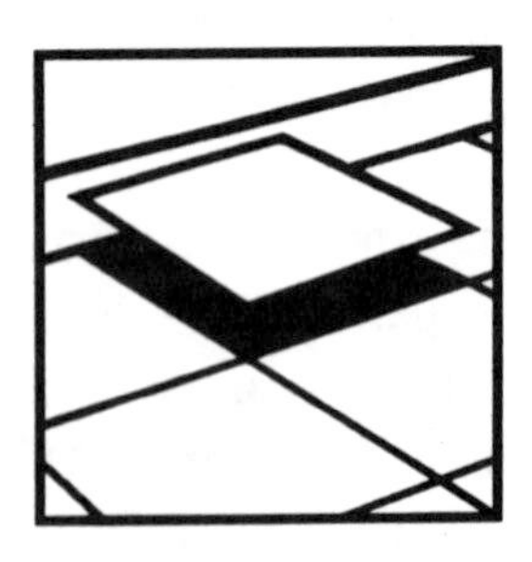

CHAPTER 2

Tile and Stone Selection

TILE AND STONE SELECTION

Are you building a new home? Restoring or re-decorating your old home? Are you an Architect or interior designer who is uncertain which tile to specify? The selection of tile and stone can be difficult, confusing and frustrating. There are hundreds of possible choices, and if you have done any shopping, I'm sure you've received an endless flow of advice. This chapter, as well as the tile selection guide on page 36, has been designed to make the selection process easier, less confusing and more enjoyable—and perhaps even to keep a few hairs from going gray.

ESTHETICS

What type of decor are you planning for the room in question? A Southwestern style may require a Mexican tile floor. If the room is to be very elegant, marble or granite may work best. When you make your selection, make sure you consider both the decor of the room and whether or not the tile complements it properly. Choose colors and styles that you will not tire of easily; tile will often last as long as the house, so you'd better be happy with your selection. If in doubt, consult an interior designer. Many tile and stone stores have designers on-staff who will be more than happy to assist you with the proper selection.

TRAFFIC AND WEAR

One of the biggest mistakes commonly made in tile selection is choosing a stone or tile that is not suited for the traffic it is to receive. Some marbles, for example, are too soft to be used on any floor that will receive high traffic. But while a busy hotel lobby may be a poor environment for some softer marbles, the same marbles may work well in a residential foyer.

How easily does the material scratch? Refer to the tile selection guide under "scratch resistance." An easy test to perform is to take a pocket knife blade and run it lightly across the tile. If the blade leaves a scratch, it will probably wear poorly in high-traffic areas. (Architects and designers should refer to ASTM C241, Abrasion Resistance).

STAIN AND ACID RESISTANCE

I will never forget a customer of mine who was a gourmet cook and installed beautiful white marble on her kitchen countertops. The marble was highly polished, very soft and *not sealed.* Needless to say, in less than a month the marble was stained every color of the rainbow and had lost its deep shine. Give careful consideration to the intended use of the tile in terms of how easily it is likely to stain or etch. If you *must* install marble on a well-used kitchen countertop, be sure to seal it properly before it becomes damaged. Refer to the tile selection guide for acid resistance and absorbency; the more absorbent the tile or stone, the more likely it will stain if not sealed. (Architects and designers should see ASTM C97, Absorption and Specific Gravity.)

COST

Unfortunately, cost is usually the deciding factor when selecting tile or stone. Yet cost can also be very misleading. A inexpensive stone or tile may fit into your budget, but if it wears easily, the cost of restoration, repair or replacement will often exceed the initial outlay. Throughly investigate the maintenance requirements of the tile you propose to purchase. Shop around and ask plenty of questions. The tile and stone markets are very competitive, so bargains can be found. Be aware, too, that ceramic and stone are available in different grades. The poorer grades may be cheaper, but will have imperfections and flaws. Examine each tile carefully before it is installed.

It would also be a good idea to spend a little more and buy spare tiles in case any need to be replaced in the future. This is especially important with marble, and ceramic since the colors and patterns you buy today may be impossible to match later on. Whatever your budget may be, do your homework and buy the best quality you can afford.

SAFETY

It is astonishing how many injuries occur each year due to slipping and falling. When choosing tile or stone, be sure it is not slippery. A highly polished granite tile installed on a shower floor is likely to be a poor choice and a dangerous slip hazard; honed, textured and flamed finishes are less slippery. There are treatments that can be applied to the surface of tile and stone to make it slip-resistant. These can be expensive ($1 to $3 per square foot), but will allow the use of tiles that are ordinarily slippery. Check with your tile and stone dealer for further information. Architects and designers should refer to ASTM test methods for coefficient of friction.

INSTALLATION AND SUBFLOOR

Who will do the installation? Many tiling projects can, in fact, be carried out successfully by the do-it-yourselfer, but certain materials such as marble should be left to the professional installer. Be sure to choose an installer who is experienced at installing the type of tile you select. A ceramic tile installer may or may not be familiar with the requirements of marble, and vice versa.

What is the condition of the subfloor—that is, the surface over which the tile is to be installed? Is there an existing floor material or tile? Many times the subfloor will have to be prepared before installation can begin. If you are tiling over an existing vinyl or tile floor, different setting materials will be needed so that proper bonding of the new tile will occur. When in doubt, ask a professional for advice.

MAINTENANCE

Maintenance is the most overlooked factor when choosing stone or tile. Just because a stone or tile costs *more* doesn't mean it's maintenance requirements are *less*. There is no such thing as a maintenance-free material; vinyl costing 20¢ per square foot requires maintenance no less than stone priced at $30 per square foot. Be sure to investigate the maintenance requirements of the tile or stone in question before you buy it. Ask to see the maintenance guide for the tile type. It would also be a good idea to check with someone you know who has the same type of tile and ask what his or her experience has been.

Remember, maintenance is a factor of *wear and use,* not of the cost of the tile.

TILE SELECTION GUIDE

TILE TYPE	COST	MAJOR USE	FINISHES AVAILABLE	ACID RESISTANT	SCRATCH RESISTANT	ABSORBANT	COLOR RANGE
MARBLE	$$-$$$	FL,FR, WL	P,H	NO	NO	1	FULL RANGE
GRANITE	$$$	FL,FR,WL	P,H,F,T	YES EX-HF	YES	2	FULL RANGE
LIMESTONE	$-$$$	FL,WL	P,H,F,	NO	NO	2-3	EARTH TONE
SLATE	$-$$	FL,WL	H,T	YES	YES	1	EARTH TONE
QUARTZITE	$$-$$$	FL	H,T	YES	YES	2-3	WHITE-PINK
SANDSTONE	$$	FL,PV	T	YES	YES	3	EARTH TONE
AGGLOMERATE	$	FL,WL	P,H	NO	NO	1	FULL RANGE
TERRAZZO	$$	FL	P,H	NO	NO	1	FULL RANGE
RIVER ROCK	$$	FL	N	YES	YES	1-2	EARTH TONE
FLAGSTONE	$-$$	FL	N	YES	YES	1	GR, BL, GR
BLUESTONE	$-$$	FL,PV	N	YES	YES	1	BL-GR
SHELL STONE	$-$$	FL,WL	P,H,N	NO	YES	3	EARTH
ONYX	$$$	FR	P	NO	NO	1	FULL RANGE
SOAPSTONE	$-$$	OR		NO	NO	2	WH
CONCRETE	$	FL	N	NO	YES	3	FULL RANGE
CULTURED MARBLE	$	FR	N	YES	NO	0	FULL RANGE
CORIAN	$$	FR	H	YES	NO	0-1	FULL RANGE
MEXICAN/TERRA COTTA	$	FL	G	YES	NO	2-3	BR, YL,RD
CERAMIC	$-$$	FL,FR,WL	G,H	YES	YES,CAN CHIP	0-1	FULL RANGE
QUARRY	$-$$	FL	H,T	YES	YES	0-1	EARTH TONE
PORCELAIN	$$-$$$	FL,WL	P,H	YES	YES	0-1	FULL RANGE
BRICK	$	FL,WL	N,G	NO	YES	3	RD,BR,TN
KEY	$-INEXPENSIV $$- MODERATE- $$$EXPENSIVE	FL-FLOORS FR-FURNITURE WL-WALLS OR-ORNAMENTAL	P-POLISHED H-HONED T-TEXTURED N-NATURAL G-GLAZED F-FLAMNED	*HYDROFLURIC ACID WILLETCH	*CAN CHIP	0-NONABSORBANT 1-LIGHTLY ABSORB 2-MODERATE ABS 3-VERY ABS	GR-GREEN G-GRAY BL-BLUE RD-RED YL-YELLOW TN-TAN WH-WHITE

SHAPES AND SIZES

MARBLE AND OTHER STONES

Marble, granite, limestone and most other popular stone is available in almost any size. The most popular stone tiles are ⅜″ thick and 12″ square. Granite and certain marble tile are also available in a ¾″ thickness and 18″ and 24″ square formats. Furniture tops are usually made from solid slabs of marble, usually ¾″ thick. These slabs are custom-cut and can be finished to almost any size and shape you might require.

CERAMIC TILE

Ceramic tile is available in a profusion of shapes and sizes. The most common are:

- Quarry tile: 6″ x 6″ to 8″ x 8″ or 18″ x 18″ square, in thicknesses of from ⅜″ to ½″.
- Pavers: generally thick, in formats ranging from 4″ x 8″ to 8 x 8″.
- Ceramic floor tile: generally available in 6″ x 6″ , 8″ x 8″ and 12″ x 12″;

some types are even produced in formats as large as 16″ x 16″ and 24″ x 24″. Thicknesses range from ⅜″ to ½″. There is also a broad variety of shapes, including squares, rectangles, hexagons and octagons.

- Ceramic wall tiles: usually ¼″ to ⅜″ thick and ranging in sizes from 2″ x 2" to 4″x 4" to 8″ x 8″.
- Mosaics: small tiles, often 1″ square. They are usually available in sheets consisting of 64 tiles held together with a cloth backing. Also produced in other shapes and sizes.

Generally speaking, you can find all types of tile in just about any desired size and shape. The formats listed here are only the ones most commonly stocked.

TILE SURFACE FINISHES

Stone and ceramic tiles are available in a variety of finishes: polished, rough-textured, smooth, etc. This section will help you identify the finish on your tile.

POLISHED

Polished marble, granite and porcelain are examples of surfaces that have a deep, reflective, mirror-like shine. A polished finish requires considerable care in order to maintain its high shine. A good general rule of thumb is *more shine, more care*. This is especially true of polished marble, since marble is much softer than granite or porcelain and is easily scratched.

GLAZED

Glazed finishes are found on ceramic tiles. Glazing is a treatment that is applied when the tile is baked. A special glazing compound, called a bisque, is applied to the ceramic after the first firing; the tile is than baked a second time to harden the glaze. The many colors and patterns found on ceramic tiles are actually contained in this glazed covering, which is extremely hard and durable. Glazes can be glossy, matte, semi-matte or textured.

UNGLAZED

Unglazed surfaces have a dull, sometimes rough appearance. Unglazed tiles typically come in earth-tone colors. Quarry tile is a popular unglazed tile.

HONED

Marble, other stones and porcelain can all be given honed finishes. A honed finish is smooth with little to no gloss. Honing is achieved by sanding the surface with abrasive to obtain a smooth, no-gloss surface. A polished stone floor may become naturally honed due to abrasion by foot traffic.

TUMBLED

Tumbled stone is made by placing tiles in a large mixer filled with abrasive particles. The resulting finish is an aged antique look with rough edges..

FLAMED OR THERMAL FINISH

A flamed or thermal finish is often applied to granite and to some limestones. It is a rough texture which is achieved by passing a very hot flame across the surface of the stone, causing pieces of the stone to pop out.

SANDBLASTED

A sandblasted finish is a rough, dull finish achieved by literally blasting the surface of the tile with sand. It is commonly given to stone surfaces. Sandblasted finishes are very porous, and require a sealer to keep clean.

HAMMERED

A hammered finish is achieved by striking the surface of the tile with a hammer and imprinting it with the pattern of the hammer's face. Hammered finishes are commonly found on granites and limestones.

NATURAL CLEFT

A natural cleft finish is often left on slate and sandstone. This is the natural finish found on the stone as it is taken from the ground.

SAWED FINISH

This is the finish produced by sawing. Commonly found on stone, it is characterized by circular saw cuts on the surface of the tile.

WAXED OR SEALED

Waxed or sealed tiles have had a coating of acrylic, natural waxes, urethane or epoxy applied to their surface. Waxes and sealers may provide gloss , water repellency and stain protection.

TEXTURED

A textured finish usually features some type of raised pattern on the surface of the tile provide to add slip-resistance qualities to the tile. It is commonly found on ceramic tiles.

CHAPTER 3

Why Do We Clean Stone & Ceramic Tile

WHY DO WE CLEAN STONE & CERAMIC TILE?

Sounds like a ridiculous question, doesn't it? Don't we clean it because it's dirty? We certainly don't want our floors or countertops looking dirty—but there are other, equally important reasons we need to clean them.

Is your tile polished—that is, does it have a shiny surface? If so, then proper cleaning will preserve that high polish. Shiny surfaces lose their shine and become dull as a result of scratching by dirt, sand and grit. This is especially true of marble.

Is your floor slippery? You certainly don't want it to be. But when dirt, dust and oil remain on a floor it may become slippery, and thus very dangerous. The thought of your head hitting a hard floor is not a pleasant one. Yet I don't think I could count the number times homeowners have told me that they *never* clean their tile floors. Neglecting proper cleaning will only result in dirt and grime so deeply imbedded that it will require an expensive refinishing process, or even replacement.

Marble & Tile

How many compliments do you receive about your stunning tile floor or your magnificent granite dining table? Marble, stone and some ceramics are symbols of achievement and success—and a direct reflection on you. There is more than one reason to keep your tile clean! Not only will proper maintenance save you a lot of money, but it will keep your investment looking new for many years. Remember, when properly cared for, tile will outlast any other building material known to man.

Let's summarize. There are five basic reasons for keeping your stone or ceramic tile clean:

1. To keep it from getting dirty.
2. To maintain its finish.
3. To prevent slipping.
4. To prolong its life.
5. To maintain the image you wish to project.

BASIC CLEANING

In this chapter we'll get down to what this book is all about—the actual care and maintenance of your marble, stone and ceramic. It is not a complicated subject. In fact, most of the mistakes that are made, and the main reason stone and ceramic get dirty, ruined and scratched, can be traced to simple neglect.

Let's suppose you've just purchased a brand-new car—and you never wash it. Soon enough the paint will fade, the body will rust, and the car will cost you a small fortune to restore. On the other hand, if you make the effort to wash it on a regular basis and throw a coat of wax on it every now and then, it will look as good as new for years. This is exactly what you need to do for your marble, stone or ceramic. Keep it clean and protected, and it will last till the kids move out—and may even be around when they move back in. The following maintenance tips will tell you what you need to know to make this possible.

LICENSE AND REGISTRATION, PLEASE?

Just as your license and registration identify you and your car to the police officer who pulls you over, knowing *what kind* of tile you have is essential to providing the proper care. Spic 'n' Span is a great floor cleaner that's been around for years, but did you know it will dull a polished marble floor? Tile X works just fine on that bathroom shower wall, but will it work as well on that marble vanity top? In fact, it will etch it terribly. It is vital to know *exactly* what type of surface you will be cleaning. The preceding chapter should help you in identification; if you are still having trouble, visit your local tile store, and take along either a photograph or a sample of the tile, if you have one. If your house is new, call the builder, ask him who installed the tile, and give that person a call. If you have marble or other stone, visit your local marble supply store, and again, take a picture with you. Do not, and I repeat, *do not* attempt to clean your tile—especially marble or stone—if you don't know for certain what it is you're dealing with. Once you do know, then, and only then, can you select the proper cleaning method.

A NIGHTMARE COMES TRUE

Mrs. Johnson just installed a brand-new black marble floor. When the installers were finished, she noticed a slight haze on the surface of the marble. Instead of bothering the tile installer who had just left, she decided to clean it herself. She went to her kitchen, grabbed her mop and bucket, added a cup of vinegar to a couple of gallons of water, and went to work. At first, the results looked promising. All the film disappeared from the new marble surface, and as long as it stayed wet, it looked really good. About an hour later, she returned to see if the stone had dried. It had, all right, but the shine was gone and the entire floor appeared white. She bent down and wiped her hand across the surface to see if the white film had returned. But this time, there was no film; the white did not come off. "Oh, my God," she thought, "what did I do?" She ran to the kitchen to make sure she had mixed vinegar and not something else in the water. No, she'd used vinegar after all, plain white vinegar.

She panicked, but finally got on the phone and called the installer—who explained that she should *never* use vinegar on polished marble. Vinegar contains acetic acid, which will dull marble; some marbles will not dull right away, but black is particularly sensitive and dulls rapidly. If Mrs. Johnson had obtained proper cleaning instructions, she would have avoided this costly nightmare.

Once you have determined what type of tile you have, the next step is to identify the finish. Which finishes are the most durable and the easiest to maintain? Here is a general guideline.

STONE

Generally, the rougher the finish, the more absorbent the tile—which means it will stain faster if not sealed and protected. On the other hand, a polished marble surface will be less absorbent, but will scuff and scratch easily.

CERAMIC TILE

A glazed surface has little or no absorbency, which makes it an excellent choice for stain resistance. However, glazed tile tends to be somewhat slippery when wet, and will chip if something sharp is dropped on it. Chips are difficult and often impossible to repair, and a brown area will remain where the glazed portion has been removed. Unglazed ceramics are more difficult to keep clean due to their greater absorbency.

PORCELAIN

Polished porcelain tile is easy to clean and does not stain easily, but it can be somewhat slippery when wet. Honed porcelain is more absorbent, and will dirty faster. Porcelain is very durable, wears well, and will not chip like other types of ceramic tile.

MEXICAN TILE

Traditional Mexican tiles are among the most difficult to care for. Classic *saltillo* pavers are very absorbent and dirty quickly. A good-quality sealer is highly recommended. *Saltillo* typically has had a stain applied to it; this stain coating eventually wears off, and is difficult or impossible to repair.

INITIAL CLEANING

The initial cleaning of a new tile or stone floor or wall should be performed by the installation contractor. If the contractor has not performed a final clean-up, or if you installed it yourself, the procedures described in this chapter are absolutely essential.

EXCESS GROUT CLEAN-UP

A new stone or tile floor may have a slight film due to dust settling from construction or an inadequate clean-up of the grout residue. It is very important that excess grout be removed before it has a chance to dry—within 24 hours for cement grout, and one hour for epoxy grouts. If excess cement grout is left on the surface for more than 24 hours, than use the following procedure:

1. Remove any large chunks of grout with a scraper or razor blade. On polished stone, take particular care not to scratch the surface.

2. Sweep or dust-mop the floor to remove all loose debris.

3. Rinse the floor several times with

plain water. Apply the water with a string mop, wrung out tightly. Avoid flooding the tile, as excessive water may cause discoloration of the grout. If too much water is applied, pick up the excess with a wrung-out string mop or wet vacuum.

4. If grout residue remains after several rinses, it will necessary to use grout-removing chemicals, as follows:

- For marble and stone: Add 3-4 oz of household ammonia to water and rinse the floor several times. A number of non-acidic grout removers are also commercially available (see the resource directory in this book).

- For glazed ceramic or porcelain: Mix a mild solution of 2-4 oz of sulfamic acid and water. Rinse the floor several times. Repeat rinsing with ammonia and water solution to remove acid residue. *Do not use any acid other than sulfamic acid.* There are several grout cleaners on the market which contain sulfamic acid; consult your local stone or tile supplier, or check with the firms and organizations listed in the resource directory.

Do not use any acids on polished marble!

EPOXY GROUT CLEANUP

Epoxy grouts are made from 100% epoxy resins. They have excellent chemical-resistance properties, and are highly recommended for tile in kitchens and baths as well as countertops and shower walls. If you decide to use an epoxy grout, make sure the person who installs it has experience working these products.

One of the biggest problems with epoxy grouts is failing to clean up the grout residue. Unlike cement-based grouts, which can sit for 24 hours, epoxy grouts need to be thoroughly cleaned *within one hour* or clean-up may prove difficult to impossible, depending on the surface type. If epoxy residue remains, the following procedure is recommended:

1. Scrape any large pieces of epoxy from the surface, using a sharp razor blade. Wetting the area first will help prevent scratching.

2. Mix a solution of hot water (the hotter the better) and several drops of dishwashing detergent (Ivory, Dove, etc.). Apply the solution to the epoxy and scrub with a green scrub-pad.

3. If the soap solution does not remove the epoxy, try wiping the surface with a clean white rag and acetone.

4. If the acetone fails the epoxy will have to be removed with a stronger solvent such as methylene chloride. Apply the solvent to the epoxy and let it stand for several minutes. Then pick up the solution with clean rags and rinse the area with plenty of water.

Caution: Solvents like methylene chloride are very dangerous to work with. If it becomes necessary to use them, I recommend calling in a professional. If you must use them yourself, be sure to have adequate ventilation and wear solvent-resistant gloves and safety glasses. Read and adhere to *all* cautions on product labels.

CLEANING AND PROTECTION DURING CONSTRUCTION

During and after the installation of a new tile floor, countertop or wall, dirt and other debris are certain to accumulate, especially if other construction is going on. Cleaning and protecting the new installation is important to prevent scratching and grinding-in of dirt.

1. Remove all equipment and tools (scaffolding, tool boxes, saw horses, ladders, etc.) from the surface. Nothing should be placed on the new surface until it is covered.

2. Remove any large chunks of concrete, grout, caulking, etc. by scraping with a sharp razor blade. Be careful not to scratch the surface.

3. Sweep or dust-mop to remove all saw dust, grit and other debris. Sweep several times, making sure to remove all loose particles.

4. Rinse the surface with cold water mixed with a small amount of neutral cleaner (see resource directory). Rinse several times, making sure the surface is clean.

A special note about flamed stone: Since flamed granite is very porous, you can expect it to be difficult to rinse. Flood the floor with clean water and immediately pick up the excess with a wet-and-dry-vacuum. Do *not* cover or seal flamed granite until it is completely dry—which may take anywhere from several days to several weeks.

5. If minor streaking remains after rinsing, buff dry with a clean white terry-cloth towel.

6. Once they are completely clean and dry, all surfaces, especially flooring, should be covered. Cover floors and countertops with kraft paper or carpet padding. Do *not* use plastic, as this will create slippery conditions cause moisture-related problems with some marbles. If necessary, tape the edges of the kraft paper with a low-contact masking tape. Do *not* use duct tape on marble, since this may cause damage.

7. If the presence of heavy objects, ladders, scaffolding or heavy construction traffic is likely to continue after covering, place plywood or masonite boards on top of the kraft paper.

8. Once all construction is complete, remove the kraft paper from the surface. Remove the tape slowly; if the tape is stubborn, wet it with a little distilled water several minutes before removing. If tape residue remains afterwards, remove it by rubbing with a white cloth and a little acetone.

9. Thoroughly dust-mop or sweep the surface to remove all debris and dust.

10. Clean the surface by rinsing with water and a neutral cleaner.

11. If the surface has developed any scratching or damage, now is the time to have it repaired. Stone can be repaired by a qualified stone restoration contractor; if ceramic tile is damaged, the installer must replace the damaged tiles .

12. It is highly recommended that a penetrating sealer be applied to all stone surfaces at this time. In the case of ceramic tile, all cement-based grout needs to be sealed with grout sealer. Mexican tile should receive a specially formulated Mexican tile sealer (see resource directory for recommendations).

CHAPTER 4

Daily Maintenance

DAILY MAINTENANCE

EQUIPMENT REQUIRED

Imagine your auto mechanic attempting a tune up on your car with nothing but a pocket knife! Of course you would expect him to have the proper tools to perform the job—and the same is true for anyone performing stone and tile maintenance. Here is a list of the tools you'll need to maintain most tile surfaces:

- Dust mop
- Damp mop
- Bucket and wringer
- Cleaning rags
- Sponge
- Neutral cleaner
- Dust pan and brush
- Spray bottle
- Vacuum cleaner (optional)

Now let's take a close look at each one of these important tools.

Dust mop. A good commercial dust mop is the most important single tool for keeping your floors looking good. For stone, use a *non-treated* dust mop; for ceramic and porcelain, use a *treated* dust mop.

A treated dust mop contains some type of dust-control chemical. You can treat the mop yourself by spraying its head with furniture polish or a commercial dust-mop spray. Treated dust mops are not recommended for stone surfaces because the oils contained in these treatments may soak into the stone and discolor it.

Do *not* store your treated dust mop with the head flat on the floor. If given the opportunity in this fashion, the treatment has a tendency to wick out of the head, making the tool ineffective.

Commercial dust mops can be purchased at most janitorial supply houses. Look in your local Yellow Pages under "janitorial supplies."

Damp mop. Two basic types of mops are available for cleaning floor surfaces: sponge mops and string mops.

Sponge mops are available in various sizes. While they are excellent mops for small areas and quick pick-up of spills, they are not generally recommended for larger areas. Since the surface area of a sponge mop is very small, it can become clogged and dirty in a very short period of time. After a few passes over the surface, the sponge mop will absorb all the dirt it can carry, and you will find yourself spreading that dirt across the floor. If you do use a sponge mop, be sure to rinse it frequently.

String mops are made in many different sizes and of various materials, including cotton, rayon and various blends of fabric. Some of these blends are designed for applying floor finishes, and are commonly called "finish mops." String mops are available with sewn or unsewn ends; we recommend a sewn end. Sewn-end mops can usually be laundered in your washing machine, while unsewn mops will shred when washed. A good quality sewn-end cotton or blend mop is best for the daily cleaning of most floor surfaces. String mops are better for larger floor surfaces, since they have a much larger surface area then sponge mops. You are much less likely to spread dirt around with a string mop.

Bucket and wringer. There are many brands of bucket and wringers on the market today; try to stay away from the cheap, poorly constructed models. A janitorial or marble supply business can steer you toward the one that's right for your purposes; generally speaking, if you have a lot of floor surface to clean, I would recommend one of the commercial models. Stay away from the metal brands and look for the ones made of plastic, since the metal types can rust and cause staining on the tile.

Cleaning rags. White terry cloth , old t-shirts and cotton cleaning rags are all excellent for picking up spills or removing streaking from a tile floor. White material is best, since it will show dirt easily, lessening the chance of using a dirty rag. Stay away from the "shop rags" sold for automotive purposes, which can easily bleed their colors onto a stone floor.

Sponges. A good-quality sponge is excellent for spot-cleaning and absorbing spills. Keep several clean sponges in a convenient location wherever accidental spills are likely to occur. Avoid sponges with a rough scrub surface on one side. This abrasive surface may scratch some of the softer marbles.

Dust pan and brush. A plastic dust pan and brush are needed to pick up dirt and debris swept up during daily cleaning. Stay away from metal dust pans, which may scratch soft stone surfaces. *Never* use a dust-pan brush to pick up wet materials; the bristles will hold the wet material and redeposit it on your floor the next time you use it, possibly causing a stain.

Vacuum cleaner. Many individuals prefer to use an electric vacuum cleaner to remove dirt and dust from tile surfaces. This is acceptable, but keep the following cautions in mind:

- Vacuum cleaners with metal wheels should not be used on stone surfaces. The metal can scratch the stone and cause rust-stain and metal-wheel marks.
- Be sure no metal parts rub across the floor's surface. Always use the soft brush attachment if the vacuum is a canister-type, and be sure any upright vacuum is equipped with a soft-bristle beater brush.
- Before using any vacuum on stone surfaces, clean the wheels and remove any sand that may be stuck to them to avoid scratching.

Neutral cleaners. Thousands of cleaners are on the market today for cleaning floors and other surfaces. Which cleaner is "best" is a matter of preference; however, it is important that any formula used to clean stone and ceramic tile have a neutral pH. A chemical's pH value indicates whether it is acid or alkaline; *neutral* pH means that it is neither, and will cause none of the damage that acid and alkaline cleaners can inflict on certain materials. If in doubt as to whether a cleaner is neutral, by all means, *ask*.

It is best to purchase a cleaner specifically made for cleaning the material in question. Marble suppliers carry stone soaps for cleaning most types of stone; neutral cleaners can be purchased from janitorial supply companies as well as marble suppliers. Be sure to follow the directions on the container carefully. Remember that too much cleaner can cause streaking.

DAILY CLEANING

To keep your stone or ceramic tile in tip-top condition, a few simple maintenance procedures are necessary. For best results, they should be followed very closely.

DUST-MOPPING

Of all the procedures used to maintain a tile surface, dust-mopping is probably the most important. Dust, dirt and grit are what cause most surfaces to scratch; if we could somehow eliminate them, scratching would cease to be a problem. Several studies have indicated that floors which are dust-mopped often stay cleaner and shinier longer.

Do dust-mop your floor daily. If traffic is heavy, or the floor is located in a commercial building, do it several times a day. Remember do not use treated dust mops on stone surfaces.

When dust-mopping, be sure to run the dust mop in one direction only. Do not move it back and forth. Think of it as pushing the dirt in only one direction. When you have accumulated enough dirt and debris, pick it up with a dust pan and brush and take the dust mop outside to shake any remaining dust. When storing the dust mop, be sure to keep its head off the ground. Hardware stores sell various hangers which are excellent for storing mops and brooms.

Designate only one dust mop for each type of surface. For example, you should use one dust mop for marble and a separate one for ceramic tile or wood floors. Do not get the dust mop wet; if the floor is wet, be sure to dry it before dust-mopping.

Purchase a good-quality machine-washable mop and keep it clean. Wash it in cold water with laundry detergent and machine-dry.

WELCOME MATS, RUGS, AND WALK-OFF MATS

Another important tool necessary for keeping dirt and debris off your floor is some type of mat, rug or what is commonly called a walk-off mat. A good-quality mat will capture dirt before someone walks on the floor. If it is placed outside your door, it is only human nature and common courtesy for people to wipe their feet before entering the house. Studies have shown that it takes approximately seven steps to remove most loose dirt from one's shoes. For this reason, I would recommend placing mats both inside and outside.

MAT AND RUG TYPES

There are hundreds of mats and rugs on the market today. Be sure to purchase one of good quality. Be careful about using rubber- or jute-backed mats or rugs on stone floors; either kind of backing can bleed into the stone, causing a stain that may be difficult or impossible to remove. A mat should be at least as wide as the doorway it serves.

Clean your mats often, daily if possible. Take them up and clean under them when you dust-mop. Be sure the floor is dry before returning them to the floor. Never place a mat down on a wet surface *or* put a wet mat on any surface.

DAMP OR WET MOPPING

All stone and ceramic tile needs to be cleaned. How often you will need to damp or wet mop will depend on the type of tile, the amount of traffic and the finish (polished, honed, glazed, etc). The following frequencies are recommended:

Marble, stone and tile (residential): Once a week
Marble, stone and tile (light commercial):Twice a week
Marble, stone and tile (heavy commercial):Daily

These are only recommendations; you will need to adjust the frequencies to suit your own conditions. For example, during a heavy rainstorm, dirt is tracked onto the tile and should be mopped up as quickly as possible.

MOPPING INSTRUCTIONS:

Fill your mop bucket about half-full (half-empty for you pessimists out there) with clean cold or warm water. *Do not use hot water.* Hot water will cause the floor to streak. Add the stone or tile cleaner (stone soap or neutral cleaner) according to the manufacturer's directions. Place your string mop in the bucket and swish the solution to mix. Wring the mop out throughly and proceed to mop the floor. Operate the mop in a figure-eight pattern; do not push it back and forth. Mop a small section, turning the mop over often. Re-dip the mop in the bucket and wring again. It is important to rinse and re-wring the mop as often as possible. The idea is to clean the floor, picking up dirt. If the mop stays dirty, you will only be pushing dirt around the floor, and when the floor dries, it will streak or appear dingy. If the floor is extremely dirty it may need to be machine-scrubbed. Certain polished stones, porcelain and some glazed ceramic tiles will tend to streak regardless of the type of cleaner and method of mopping; if this proves to be a problem, dry the tile with a clean terry cloth towel, or, if possible, machine-buff with a white pad. The small two-headed buffers sold in department stores are excellent for buffing small areas.

MACHINE-SCRUBBING

Certain surfaces such as flamed granite, quarry tile and most rough-textured tiles are very difficult to mop. They tend to be very absorbent, soaking up water as fast as you can put it down, and some will actually shred your mop head. If this is a problem, machine-scrubbing may be necessary.

There are several types of floor buffers that can be used. The standard janitorial buffer is ideal, but if you have never operated one of these machines you will find it difficult to control and may end up causing more harm than good.

For the home owner or inexperienced operator, several smaller two- and three-headed machines are available on the market which are easy to operate and relatively inexpensive. They can be purchased in department stores or from tile dealers. On the other hand, these small machines, though excellent for the homeowner, are not intended for daily commercial use. They can also double as floor polishers for polished stone and wood flooring. Most rental yards, by the way, carry both small and commercial floor buffers, check your Yellow Pages.

MACHINE-SCRUBBING INSTRUCTIONS

Prepare a solution of tile cleaner as outlined in the mopping instructions above. Instead of wringing the mop, dip it in the bucket and mop a small section of the floor. You will notice that the mop will deposit quite a bit of water on the surface. Place the mop back in the bucket and scrub the area with the floor machine. Work the solution back and forth, making sure to clean each area throughly. The floor scrubber should be equipped with soft nylon brushes or white pads. Pick up the solution with a wet vacuum or mop it up with a string mop. Do *not* use the same mop for picking up the dirty solution; use a separate mop *and* bucket/wringer. After the whole floor has been scrubbed, it may be necessary to mop the surface again to remove any light film or puddled water.

CHAPTER 5

Heavy Duty Cleaning

HEAVY DUTY CLEANING

This chapter is for all you bad boys and girls who have neglected the daily maintenance of your tile floor. Heavy-duty cleaning is required when too much dirt and grime have accumulated and daily cleaning doesn't work any more. If you find it necessary to give your floor a heavy cleaning more than once a year (I'm speaking of residential applications here), see the chapter on sealing; a properly sealed floor will be much easier to clean and will stay cleaner longer.

All right, let's assume that for whatever reason, your floor is a mess and needs some help. Here is how we get stubborn floors clean.

HEAVY-DUTY CLEANERS:

Heavy-duty cleaners are usually strong, highly alkaline formulas such as floor wax strippers, degreasers, commercial floor cleaners, etc. They are typically not pH neutral, and thus require more care in handling than a neutral cleaner. Many heavy-duty cleaners are acidic, and these should not be used on polished marble, limestone and even some glazed ceramics.

Marble & Tile

Be sure to follow the manufacturer's instructions precisely when using these cleaners, and *do not mix them with any other cleaners.*

MIXING

If a little is good, then more is better—*not!* This is the biggest mistake made by homeowners and professionals. *Never* mix more than what is suggested on the product's label. The suggested dosage has been throughly tested; problems such as streaking, chemical reactions and even destruction of the tile can occur if too much of the product is used. Some chemicals only react with a certain quantity of water, and if too much chemical is used, no reaction will take place—and not surprisingly, the product will not do what it's suppose to do.

HEAVY-DUTY CLEANING INSTRUCTIONS:

Most heavy-duty cleaning will require the use of a floor machine. If one is not available and you don't mind using a lot of elbow grease, it's possible to do the job with a soft-bristle scrub brush. If you do use a scrub brush, I recommend purchasing one with a long handle—it's a lot easier to work standing up than on your hands and knees. The following procedure works reasonably well for most heavy-duty cleaning situations; if the floor has a coating of wax, however, proceed directly to the section on stripping.

1. Prepare a solution of cleaner as directed on the label of the container. Use cold to warm water, unless otherwise stated on the label. Remember to mix exactly to directions; more is *not* better.

2. Apply the solution to a small section of the floor, taking care not to splash any of the solution on adjacent surfaces. Some heavy duty cleaners will strip paint and varnish off woodwork, etc. Allow the solution to soak on the floor for several minutes, but do not let it dry. If it begins to dry, add more solution or plain water to keep it wet.

3. After the solution has had time to work on the dirt, either machine-scrub or get out the old scrub brush. If using a buffing machine, equip it with a nylon brush or a red or tan pad. Be especially careful with polished marble and test the pad or brush to make sure it will not scratch.

4. Pick up the solution and rinse several times. Rinsing is extremely important, as some heavy duty cleaners will leave behind a powdery residue which can be difficult to remove.

5. Examine the area carefully to make sure it is completely clean. If not, clean it again.

6. If a coating or sealer is to be applied, make sure floor is dry. Placing several fans on the floor will increase the drying time.

WAX REMOVAL AND STRIPPING

Suppose your tile floors have been covered with a wax or some other type of coating. How do you remove this coating, and what chemicals should be used? The procedure used to strip a tile floor is relatively simple, and is outlined below. But before you run out and buy a chemical floor stripper, it is important, as always, to find out what is *on* the floor.

WHAT'S ON MY FLOOR?

Today's technology has delivered hundreds of different types of floor coatings—natural and synthetic waxes, acrylics, thermoplastics, polyurethanes, epoxies, etc. To choose the proper chemical for removing them, we must know what type of coating we are dealing with. I suggest the following procedure:

1. If you already know what coating is on the floor, contact the product's manufacturer for removal instructions.

2. If you have no idea what is on the floor, perform the following test. Mix one

cup of household ammonia in one gallon of warm water; pour a small amount of this mixture on the floor and agitate with a soft scrub brush. Pick up the solution with a wet vacuum or a dry rag. Examine the area; if you have removed the coating, you will need to use a commercial wax stripper to remove the remaining wax. Visit your local janitorial supply house and ask for an alkaline acrylic finish stripper. If the coating has *not* been removed with the ammonia, there is a good possibility that the finish is urethane- or epoxy-based. What this means is that some very strong solvents will be needed to remove the finish. At this point, it might be wise to call in a professional.

STRIPPING PROCEDURE

To strip a tile floor with commercial alkaline strippers, use the following procedure. A janitorial floor-buffing machine and a wet vacuum are highly recommended. Before setting up to strip the tile be sure to remove all furniture and protect painted surfaces such as baseboards with plastic drape.

1. Mix the stripper solution in a separate bucket, following the instructions on the label.

2. Apply the mixed solution with a string mop to one small section at a time. Do not apply more than can be scrubbed effectively. Let the stripper sit for several minutes; this will allow the chemicals to break up the coating. If the stripper begins to dry, add more of the solution to keep it wet.

3. Scrub the floor using a 175-rpm standard buffing machine equipped with a black stripping pad or stripping brush (pads and brushes are available at most janitorial and rental supply houses). Continue scrubbing until the coating breaks up. Some soft marbles may be scratched by stripping pads and brushes; always test a small area first before proceeding with the entire project. If the pad scratches, your janitorial supply house can recommend a softer pad or brush.

4. Pick up the solution with a wet vacuum and rinse immediately. Use a separate string mop, bucket and wringer for your rinse water.

5. Examine the area thoroughly. If any of the coating is still present, re-strip.

When you're through, be sure to rinse the floor thoroughly; most alkaline

strippers have a tendency to leave a film. Adding several ounces of a good neutral cleaner or stone soap to the rinse water will help neutralize the stripping solution.
If a coating is to be applied, proceed to chapter 8 of this book. If a penetrating sealer is to be used, make sure the floor is thoroughly dry. Allow at least 24 hours before sealer application.

CHAPTER 6

Restoration & Polishing

RESTORATION & POLISHING

So your marble is dull, scratched and in major need of help! Out come the Yellow Pages; you call several marble refinishers and set up a few appointments. *How hard can it be?* You'll get a few estimates, check some references and select a professional to do the job.

And then the fun begins.

The *first* professional tells you that your floor needs to be ground flat in order to be fixed properly.

The *second* professional tells you he can hone and polish it—no grinding required.

The *third* professional tells you he only needs to *recrystallize* the marble to make it look like new.

Now that you're totally confused, how do you determine who's right? What's the difference between grinding and honing? What is polishing and "recrystallization?" This chapter will arm you with enough knowledge of these terms to help you ask the right questions. Once you are familiar with the processes used to restore marble and stone, we will provide some guidelines and suggest some questions to ask the contractor.

WHY DOES STONE SHINE?

When stone becomes dull and scratched, it obviously loses its shine and luster. At this time the stone needs to be refinished and polished to restore the shine it had originally. Why does stone shine, and how can a lost shine be recovered?

All stone is taken from the earth in the form of raw *blocks.* Explosives, large saws and specialized equipment are used to extract the stone from the earth. The stone blocks are then cut into thinner, more easily handled pieces called *slabs.* The slab itself is then processed, depending on the intended use of the stone. It may be given a high shine and shipped to a marble fabricator, who will ultimately turn it into a table, vanity top or whatever; or it may be transformed by some *very* expensive and sophisticated equipment into tiles for installation on floors or walls.

The deep shine we see on polished stone is achieved by rubbing the stone with a series of abrasive materials. The process is very similar to sanding a piece of wood. The stone is rubbed with a coarse abrasive grit, followed by finer and finer grits until the stone becomes smooth. The scratches left behind from one grit are removed by the next, creating finer and finer scratches. The process continues until the scratches are microscopic. The shine on the stone is achieved by abrading the surface to the point at which it becomes extremely smooth and starts to develop some reflectivity. The shine on the stone is thus a product of optics. In Diagram A, we see a rough or scratched piece of stone. When light is reflected from the surface the light rays scatter, producing a dull, flat appearance to the human eye. In Diagram B, the stone has been smoothed to such a degree that when light is reflected from the surface, the light rays return in a parallel pattern, producing a reflection to your eyes. This same optical property can be observed on a pond. When the wind is blowing and the surface of the pond is wavy, it becomes difficult to see a reflection; when the air is still and the pond is calm, a deep reflection can be observed. So in order to achieve a deep shine on your stone, all that really needs to be done is to smooth it until it shines.

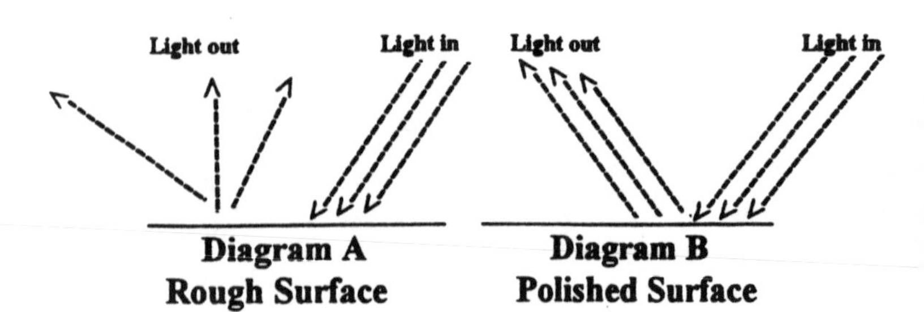

Diagram A
Rough Surface

Diagram B
Polished Surface

Sounds simple enough, doesn't it? Unfortunately, the techniques employed to achieve this degree of smoothness require special knowledge and training. This is *not* friendly territory for the do-it-yourselfer.

To help clear up the confusion, let's define some terms, then move on the all-important issue of selecting a stone/tile professional.

GRINDING

Grinding is the process by which the surface is aggressively sanded to remove large stocks of the stone. This process is usually recommended when stone tiles are uneven. *Lippage* is the term given to uneven tiles that are set higher than one another. Grinding is recommended when the lippage exceeds ⅛″ inch or if one desires to have a completely flat floor.

There are some very good reasons for grinding a stone floor flat. A flat floor is easier to maintain; since there will be no lips where dirt can accumulate. The grinding process, if performed correctly, will also eliminate depressed grout joints—the grout will be even with the tile's surface so that dirt and grime can't accumulate.

A completely flat floor eliminates all unevenness, giving the floor the illusion of being monolithic (one piece).

Note: a stone floor does not necessarily have to be ground to remove scratching. A skilled craftsmen can repair it without grinding.

Just as there are several good reasons for grinding, there are also some disadvantages. Grinding is very time-consuming and expensive; with some hard stones, like granite, it can take an entire day to grind 50 square feet. The grinding process is also very messy. Copious amounts of water are needed to grind a stone floor and produces a heavy slurry of stone and water. If adjacent areas such as carpet, wallpaper, baseboards, etc., are not protected properly, water damage may occur.

Before deciding on grinding, all the above considerations must be carefully weighed. Discuss the options with the stone specialist.

HONING

Honing is the process of smoothing the stone with the use of abrasives. Although not as aggressive as grinding, it does require the use of water, and can also be quite messy. Honing is performed to remove scratches, and will not remove lippage (uneven tiles). It can, however, round the edges of the stone, giving a smoother finish to the edge. The honing process is usually achieved with the use of diamond abrasives, although some contractors prefer silicon-carbide bricks or screens. Which abrasive is used is not as important as the skill level of the craftsmen. Honing can leave a stone floor with very little shine, although some stones will acquire a satin-like luster at very high hones.

You may hear the contractor talk about *grit sizes* when discussing the honing-and-grinding process. The following table will serve as a guide to grit sizes. The lower the number, the more aggressive the grit. Generally, *grinding* is what takes place using any grit of 60 or below; *honing* begins at 120 and proceeds upwards. A skilled craftsman will generally stop at a 400 or 600 on marble before *polishing*. With granite, it is usually necessary to proceed through to the highest grit. Some craftsmen may choose to polish with diamond abrasives to the highest grit, producing a very high polish, while others may choose to switch from a diamond to a powdered abrasive (see next section). Whichever method is chosen, the final result is what counts.

GRIT SIZE TABLE

The following table list some of the most common grit sizes used in the stone industry. The lower the number, the more aggressive the grit.

16	
24	Grinding
36	
60	
120	
220	
400	
600	Honing/Polishing
800	
1,800	
2,000	
3,000	
3,500	
5,000	
8,500	

POLISHING

As previously discussed, the high shine observed on stone is the result of smoothing it with fine abrasives. Most craftsmen will use diamond abrasives to hone the stone, then switch to a powdered abrasive to achieve the final polish. Powdered abrasives contain superfine crystals of aluminum oxide or tin oxide. These powders are usually white, but can be yellow, brown gray or black.

The abrasive powder is worked into the stone with a floor machine using water and cloth or polyester fiber pads. The powder is worked into a slurry until a polish is achieved. The craftsman removes the slurry with a wet-vac or mop and rinses the floor to remove excess powder. It's a relatively simple procedure, but it requires a good deal of practice for several reasons. Many polishing powders contain a compound known as oxalic acid, which is used to speed the polishing process, and if too much powder is used, the stone can burn. A burned floor has a characteristic dimpled appearance; the stone will have a molten, plastic shine. This burned appearance is commonly called "orange peel," for reasons that are obvious to anyone who sees it. If the craftsmen orange-peels the floor, he will have to re-hone the floor to remove it. On the other hand, if too *little* powder is used, the final polish may not be achieved. A good craftsmen will be familiar with the powder polishing technique.

RECRYSTALLIZATION

The term *recrystallization* has entered the language of the marble polishing field to describe a process used to maintain a shine on marble surfaces. Recrystallization can also be called "vitrification" or *incorrectly* called "marble polishing." The procedure has been used in the United States since the 1970s, and has generated some controversy among the experts. Before we discuss the pros and cons of this process, we need to take a look at what it actually entails.

The recrystallization process consists of spraying a fluid onto the marble floor and buffing it in with steel wool under a standard buffing machine. The steel wool generates heat through abrasion and the chemical reacts with the marble, producing a new compound on the surface of the stone.

Sounds simple enough, so why the controversy? Proponents of the process claim the new compound formed protects the surface of the stone, adds shine and may even harden the stone, increasing its wear. Opponents of the process claim that the new compound that is formed blocks the stone's ability to "breathe," traps moisture and causes the stone to rot.

Both sides have put forward convincing arguments, but at this time, the jury's still out. If you opt for recrystallization, it is extremely important that the process be carried out only by trained craftsmen who are thoroughly familiar with it.

One additional note: the recrystallization process can only be applied to marble and limestone. Recrystallization cannot be used on granite, quartzite and sandstone.

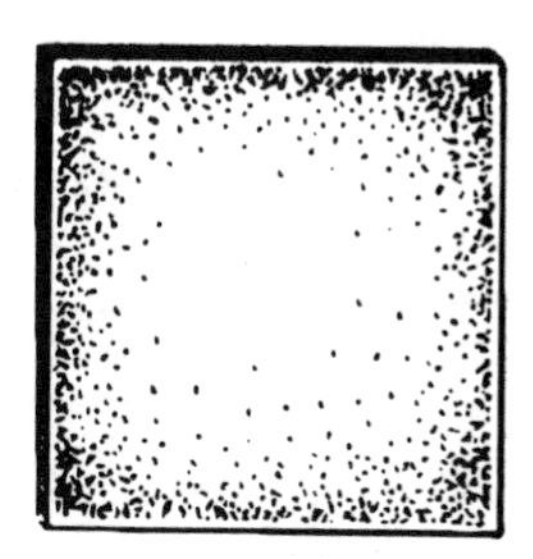

WHICH PROCESS TO CHOOSE

It is very difficult for an untrained eye to evaluate your marble floor. Contact a reputable stone refinisher, ask the right questions and check references. The remainder of this chapter will offer guidelines to help you choose the right contractor.

HIRING A PROFESSIONAL

Hiring a professional stone contractor or tile setter can be difficult. A careful reading of this book should give you enough background to know what a professional is talking about. Here are some points you should definitely cover before arriving at any decision:

1. Before calling any contractor, ask around. Have any of your neighbors had their marble/tile done recently? Who did it, and were they happy? Call your local stone/tile suppliers and ask whom they'd recommend for this type of work. Some of the stone/tile Associations listed in this book's question-and-answer section may provide recommendations. And of course there are always the Yellow Pages.

2. Once you have located several companies, schedule appointments to receive estimates. Almost every contractor I know will perform a free estimate. Be sure you are there for the scheduled time; it can be very frustrating for a contractor to arrive for any estimate, only to find no one home. On the other hand, if the contractor fails to show for the scheduled appointment without at least calling, he obviously isn't interested in your project.

3. When the contractor arrives, explain what your concerns are and what you are trying to achieve. After all, you live with the floor every day; the contractor is seeing it for the first time. Give the contractor as much information as possible. What do you use to clean the floor? Has the floor been polished? Is there any wax or coating on the floor? If it's a new installation, the contractor will also need to know what materials are on the floor now. Any information will help him decide how to fix the problem or whether to install a new floor.

4. Once the contractor has determined what is needed, ask him to explain the procedure he intends to use. Are there other options? Reread the section on restoration and polishing and ask specific

questions. What polishing process will be used, etc.? A competent contractor should be more than happy to answer any question you may have.

5. A word on negotiating price: Among contractors, as in any occupation, personalities vary widely. Some contractors will negotiate; others will stick to their guns—although if you mention that you are getting two additional estimates, even a stubborn contractor may sharpen his pencil. Above all, make sure you're comparing apples to apples. If one contractor is only going to polish and the other is going to grind, the difference in price will be considerable. If new tile is to be installed, will it be mud-set or thin-set?

6. If possible, obtain a demo or sample. Ask if a free demo can be performed; have it performed in a representative area. This will indicate what the final job will be like. Be reasonable, however; don't expect a contractor to perform a demo if the job is too small.

7. Ask for references—and check them. Many contractors in all fields have references, but you'd be surprised how rarely they are actually checked. Call at least three and ask if the contractor did a good job.

Were there any problems and did he correct them? Where his employees professional?

8. Does the contractor carry insurance? Ask for proof. Have him show you a certificate of insurance, or, if the job is large enough, have his insurance company send you one. Be sure he carries liability and workers' compensation insurance. Any reputable company will carry both.

9. Once you choose a contractor, schedule the job. Don't be surprised if the contractor is booked for several weeks. Be patient; a *good* contractor will be busy, and you will have to wait your turn. If you absolutely *must* have it done now, ask him if he'll book you if he gets a cancellation.

10. Gut feeling—are you comfortable with the contractor? This is much more important than you might think.

Even the best contractors can make mistakes. The difference between a good contractor and a bad one is the *willingness to correct those mistakes*.

CHAPTER 7

Slip Resistance

SLIP RESISTANCE

Is the floor tile you specified or purchased slippery? If it is, is there a treatment that can be applied to make it slip-resistance. If your a Designer or Architect , is there a slip resistance standard that a floor tile must meet? If so, who regulates these standards? These are just a few of the questions that need to be addressed when selecting floor tile or any walking surface. There are coatings and treatments that are available that can be applied to floor tile to make it slip-resistant. Tiles can also be purchased that are less slippery than others. What are these treatments and how are they applied? What can a designer do to make sure the floor tile is safe and people don't slip and fall? This chapter will address these issues as they apply today.

COF-Coefficient of Friction

Slip resistance is measured by the ratio of forces required to move one surface over the other under a given vertical force. In other words, it takes two surfaces to determine slip resistance. The floor tile is one surface and the bottom of ones shoe is the other surface. This ratio is what we call the coefficient of friction(COF).

COF can be measured in two different ways and can cause confusion amongst those unfamiliar with the science of slip resistance. When the COF is measured from a resting position it is called the "Static COF". When it is measured when the surfaces are in relative motion it is called the "Dynamic COF". The dynamic COF is very difficult to measure and almost all portable and laboratory meters measure only the static COF. It is important to know this difference since you will see both measurements in the literature. Most measuring devices(Slip meters) will refer to the static COF. The measurements you will find in the literature and those discussed here will be the Static COF. A COF of 0.5 is considered to be a slip resistance surface. The higher the COF the less slippery the surface. It is possible to have too high a COF. In other words the surface can be too slip resistance and an individual would find it difficult to walk on.

How to Measure Slip Resistance

There are basically two types of machines that can measure static COF, Permanent laboratory models and portable field models commonly refer to as pull meters. The most popular and widely accepted laboratory slip meter is the James Machine. The James machine uses an 80 pound weight that is applied through an arm to a leather shoe placed on a panel. The panel and the leather shoe are moved horizontally. The distance the panel moves before slipping is measured and recorded on a chart and is the coefficient of friction. The James machine was invented in 1940 and was the machine which established the 0.5 COF as the minimum for slip resistance. This standard was accepted in 1953 by the Federal Trade Commission . Many still consider the James machine as the only true slip tester.

There are also many portable slip meters on the market which claim to measure the static COF. The American Society for Testing and Materials(ASTM) recognizes several portable meters. When purchasing these meters make sure they comply with ASTM C-1028 which is the recognized slip test for tile flooring. For further information on slip meters contact ASTM at the following address: 1916 Race Street, Philadelphia, PA 19103-1187

The ADA and COF

On July 26, 1991 the Americans with Disability Act(ADA) was passed. This document covers recommendations and requirements for designing and retrofitting public spaces for the disabled. This document covers many aspects of construction that all designers and architects should be familiar with. One of these recommendation deals with Coefficient of Friction. If one examines the ADA document it can be rather confusing as to the exact requirements of COF and to whom or where they apply. I placed a call to the ADA office and spoke with several representatives and they all told me that this was a gray area and that the COF sited was only a recommendation and not a requirement. Good answer but, that doesn't help the architect , designer or building owner who may become prey to an accidental or even deliberate falls. So what can you do as a designer, building owner/manager or flooring contractor to make sure your tile floor is slip-resistant.

1. Ask the tile supplier or manufacturer for slip testing data. Almost all the tile manufacturers have this data. It may be difficult to obtain slip readings for stone tile. If this is the case there are many independent testing labs that can perform an ASTM COF test. This is highly recommended if the COF of the surface you will be using is not readily available. The COF should have a static COF of at least 0.5. Important note; Certain

county and city governments have their own slip resistance codes. It is important to check with the local government where you intend to use the tile and/or flooring.

2. Avoid highly polished finishes. Polished marble, granite and glazed ceramics and porcelain can be slippery. If you must use these finishes and they do not meet the COF requirements than you will need to specify a slip resistance treatment or coating.

3. Specify proper maintenance for the selected tile. In many instances the tile floor is not slippery when installed new but lack of proper maintenance, build up of dirt and grime etc. will contribute greatly to the slip resistance of a floor. If you don't know how to maintain the floor then ask the manufacturer of the tile or consult an expert in floor maintenance.

How to make a slippery floor safe

There are many treatments that can be purchased and applied to a tile surface so that it meets the ADA recommendations of 0.5 COF. These treatments fall into two categories: Coatings or treatments which chemically or physically alter the surface of the tile.

Coatings:
Coatings can be waxes, acrylics or other commercially available products. The coating places a thin layer of material on the surface of the tile. The tile will than be as slippery as the coating itself. Warning: not all coatings provide slip resistance. In fact some coatings will make a floor more slippery. Before specifying a coating contact the manufacturer of the coating and ask for slip resistance test data. Many of the coating manufacturers are very familiar with slip resistance. Make sure the coating can be used on the tile surface you are using. Certain coatings will not adhere to polished stone or porcelain and require coatings specified for these surfaces.

Treatments:
There are now available special treatments that can be applied to the surface of tile to render it slip resistant. These treatments are primary hydrofluoric acid. The acid attacks the surface of the tile and creates microscopic holes. This is what is typically called etching of the surface. This process works effectively on may surfaces but can decrease the serve life of the tile. Once the surface is treated with this method maintenance will increase. Since these treatments contain a very dangerous acid, it should only be applied by trained individuals.

Marble & Tile

Contact your local tile supply store for recommended contractors.

The issue of slip-resistance is of major concern in the US. Lawsuits are on the increase as con artists continue their search for the big payoff. Large hotels , banks and other big corporate building owners are their main targets but they are also targeting the small business. The corner food store, the local gas station are not exempt from these flim flam artists.
This is not to say that there are not times when a floor surface is not unsafe. Many factors contribute to the slipperiness of the floor. Water, grease, oil and debris scattered on the floor all can contribute to slipperiness. The competent architect and designer cannot control what happens to the floor after it's installed but he/she can get it started properly.

How to Minimize Slip/Fall Accidents

Although it will be impossible to prevent all slip/fall accidents there are several precautionary procedures that building owners, cleaning companies and others can take to minimize their risk. The following are some suggestions and is not intended to replace legal advise if an accident occurs:

Marble & Tile

1. Pay attention to areas where water and/or spills occur. A walk off mat should be placed inside the entrance of doors during a rain storm. When floor tile gets wet, the COF may decrease causing a fall. Mats should be placed prior to the first drop of rain. Pay attention to areas where food is served or carried. Foods and drinks can create an ice like condition of the floor and any spills should be picked up as soon as they occur.

2. If the floor tile is maintained by stripping and waxing, this procedure should be done at night when there is no one around to fall. Daily wet mopping should also be performed at night during off hours.

3. Always place wet floor signs in all areas you may be working in. This applies for all times of the day or night. It is also a good idea to train your floor cleaning personnel to warn people who may walk across the floor that it might be slippery.

4. Keep accurate records. It is surprising how many cleaning companies fail to keep any record of maintenance on a tile surface. Accurate record keeping says that you are responsible and that you generally care about safety. Include in your records the following information:

* Name brands of all products used on the floor.

* Procedures that are performed on the floor and how often. Be specific.

* Who performed these procedures?

It is a good idea to keep a daily log of the maintenance procedure. Designate one individual to keep track of the log and have him perform routine inspections of the floor and record what he found during his inspections. If a legal suit is filed, this information will be quite helpful in proving you are competent and hopefully not at fault.

5. Get slip/fall insurance. Every building owner should carry slip/fall insurance. Some of the products used on the floor will also carry slip/fall insurance. This insurance is designed to protect the coating manufacturer and should not rely entirely on theirs. Get your own policy.

What to do when someone falls:
Oh my, someone just slipped and fell, what should you do?

1. Treat the victim with kindness and courtesy. Do not administer medical treatment unless you happen to be qualified

to. If necessary call an ambulance. Also, offer to call their family. Show extreme care and concern.

2. Look at the victims clothing and especially his/her shoes. Take note of how worn they are and the overall condition. Once the victim leaves, record this information in your log or on a separate report. This is important, remember it takes two surfaces to slip, if the victim is wearing worn shoes part of the blame may be placed on their neglect.

3. Look carefully at the area where the victim fell. Is there any water, grease, a banana peel? Record in the log anything you find.

4. Did anyone else see the fall? If so take a statement from them. Record this in a special report and get their name, address and phone number.

5. Record the time and location of the fall? Pay special attention to the weather conditions. Was it raining, snowy or what?

6. Did you notice how the victim walked? Did he/she stumble or appear as if they were under the influence of alcohol or drugs? If possible record their behavior and how they

walked or ran before the fall.

7. If you have a camera, take pictures of the victim and the area where they fell.

8. Fill out an accident report and do not leave out any detail. Use a separate sheet of paper if necessary.

Following the above suggestions will help considerably when involved in a slip/fall case. Your attorney will love you for it.

One Final Word

The legislation and rules concerning COF and slip/fall can and probably will change. It is a good idea to consult with an expert in slip/fall and to keep up to date with any new rule, law or recommendation that may sneak up.

CHAPTER 8

Stain Removal & Protection

STAIN REMOVAL & PROTECTION

Oh, those nasty stains! How do you go about removing that red stain from the fruit drink your son spilled on the floor? What do you use to remove black marker ink from your white marble floor? Does lipstick stain? The list can go on and on.

Removing stains from marble, granite and ceramic tile can prove difficult. There are, however, several precautionary measures you can take:

• *Any spill should be cleaned as soon as possible.* Blot spills with a paper towel or clean rag. At this stage, it is important *only to blot;* wiping a spill may spread it over a larger area, making a larger mess. Use only cold water and stone soap or a neutral cleaner. Rinse the area several times. If a stain is still present, a chemical poultice may have to be applied.

• *Avoid using chemicals of any kind* until you know which chemical cleaner to use (see accompanying chart). Certain chemicals will react with the spilled material, and could make the stain permanent.

Marble & Tile

Marble, granite and certain ceramic tile are porous materials. If not thoroughly sealed, they will stain. The only way a stain can be removed is to literally pull it out of the stone or ceramic with both a chemical and material that will absorb the stain. This chemical absorbent-material combination is what we call a *poultice*.

Poultices are commonly powder or cloth materials that can be mixed with a chemical and placed on top of the stain. Refer to the table below for some of the more common poultice materials. Clays and diatomaceous earth are safe and readily available, but do *not* use whiting or clays containing iron with an acidic chemical; iron will react with the acid, and may cause rust staining. It is best to purchase powders that are designed specifically for stone and tile. Consult a stone restoration specialist or your stone supplier if in doubt.

Poultice materials:
Paper towels
Cotton balls
Gauze pads
Clays such as attapulgite, kaolin, fuller's earth
Talc
Chalk (whiting)
Sepiolite
Diatomaceous earth
Methyl cellulose
Flour
Saw dust

How to apply a poultice

Before you attempt to remove a stain, it is extremely important to know what has caused it. If you don't know, I would recommend that you consult a stone specialist, or refer to my book on stain removal for a detailed description of the procedure.

To apply a poultice, take the following steps:

1. Clean the stained area with water

and stone soap. Remember to blot rather than wipe.

2. Pre-wet the stained area with a little water. Distilled water is recommended.

3. Refer to the chart and determine which chemical to use for the stain.

4. Mix the poultice material with the selected chemical. Mix until a thick peanut-butter paste consistency is obtained.

5. Apply the paste to the stained area, overlapping the stain by at least ¼". Do not make the application too thick, or it will take a long time to dry.

6. Cover the paste with a plastic sandwich bag or food wrap. Tape the plastic using a low-contact tape.

7. Allow the paste to sit for 12–24 hours.

8. Remove the plastic cover and check to see if the paste has dried. If it has not, allow it to sit uncovered until thoroughly dry.

9. Once it is dry, remove the paste by scraping and rinse the area.

10. Examine the stain. If it still remains, but is somewhat lighter, re-poultice until it is gone. If the stain refuses to disappear completely, it is time to give up, replace the tile or call a stone specialist.

Stain removal can be very difficult, and care must be taken when using a poultice. The following stain removal chart represents only a small sampling of the chemicals that are available for stain removal; a complete guide can be found in my book on the subject.

Stain Removal Chart

Stain:	Poultice material:
Ink	Baking soda or hydrogen peroxide
Cooking oil	Baking soda or ammonia
Tea and Coffee	Hydrogen peroxide
Rust	Glycerine and sodium citrate
Mildew/ algae	Household bleach
Copper	Ammonia
Paint	Commercial paint remover
Wine (red)	Hydrogen peroxide
Fruit	Hydrogen peroxide
Water spots	None; rub with #0000 steel wool

SEALERS, WAXES AND IMPREGNATORS

There are hundreds of brands of waxes, and sealers on the market today. Some are good, while others are not so good. Before you can choose the best product to use, you first need to determine whether the tile needs a wax or sealer. Some porcelain and ceramic tiles have very low absorption, and will not need any wax or sealer; some, such as Mexican and stone tiles, can be very porous, and will definitely require some type of a sealer.

Waxes and sealers serve two purposes: they provide protection to the surface against staining, and they may improve the shine. To make matters more confusing, some products on the market offer *only* protection, while others only add shine. All sealers, however, fall into one of two general categories. They are either *coatings* or *penetrating sealers*—often called *impregnators.*

COATINGS

Coatings are sealers that place a sacrificial coating on the surface of the stone or tile to act as a barrier to prevent water, oil and dirt from penetrating and staining it. Coatings are usually described as waxes, sealers, floor finishes, polishes, etc. Most of the floor finishes you can purchase at the grocery store fall into the coating category. Whatever they are called, we can further divide most coatings into one of two categories: *strippable* or *permanent.*

Strippable coatings are designed to be easily removed from the surface of the tile. They are usually made of polymers, including acrylics, styrene, polyethylene and hundreds of other polymer types. Most are water-based, which means they will also dissolve in water. This characteristic makes it easy to remove them with strippers. There are hundreds of strippable-coating formulas. Some are designed for vinyl tile, others designed for Mexican, stone, quarry and ceramic tile. If in doubt, always ask what type of surface is the coating designed for.

Permanent coatings are very difficult to remove. They are made of urethane or epoxies, which require special solvents for removal. There are some very rare situations in which permanent coatings should be used. Polyurethanes are used on wood flooring, but should not be used on stone, ceramic or other hard-surface flooring materials.

PENETRATING SEALERS

As their name implies, impregnators or penetrating sealers are designed to penetrate below the surface of the stone or tile and either deposit solid particles in the pores of the stone or tile, or coat the individual minerals below the surface. Penetrating sealers work by restricting water, oil and dirt from entering the stone or tile.

There are many brands of penetrating sealers/impregnators on the market. They can be solvent-based or water-based, and generally contain silicone, siloxane, silane, methyl silicate or other silicone derivatives. Occasionally you will hear them called *silicone sealers*. Like coatings, penetrating sealers also fall into two categories: *water-repelling* (hydrophobic) and *oil-repelling* (oleophobic).

Water-repelling penetrating sealers are designed to repel only water and the contaminants carried by water. Fruit drinks, coffee, tea and soda are examples of water-carried products that would be repelled by water-repellant penetrating sealers.

Oil-repellant penetrating sealers will repel water as well as oil-based contaminates. Cooking oil, grease, butter and body oils are all examples of oil-carried products that would be repelled by oil repelling penetrating sealers.

A word of caution: most penetrating sealers are designed to be either water- or oil-*resistant*, not water- or oil-*proof*. If a liquid is left on the stone or tile long enough, it will eventually penetrate and stain. Penetrating sealers are designed to give you *time* to clean up the mess before staining occurs. Nor are they designed to prevent acid etching; lemon, vinegar , tomato and other acidic foods and drinks will etch the surface of most marble and limestone, whether sealed or unsealed.

WHICH SEALER DO YOU NEED?

As you can see by now, deciding which sealer to use can be a confusing process. The best way to proceed is to ask yourself the following questions:

1. What type of stone or tile do I have? Some stones are more porous than others, while certain tiles will have very little porosity. Granite, for example, is generally more porous than marble, while porcelain tile will be less porous than stone. The more absorbent the stone or tile, the more difficult it will be to seal. To determine the relative porosity of your material, perform the following test. Place several drops of water on top of the stone or tile and use a stopwatch to tell how long it takes for the water to completely disappear and soak into the surface. If the water disappears in less than one minute, consider it *very porous*. If it takes under three minutes, consider it *porous*. If it takes three to four minutes or even longer, call it *slightly porous*. Of course, if the water does not disappear at all, sealing may not be necessary.

2. What kind of texture does the tile or stone have? A highly polished surface will generally be less porous than a honed finish; a honed finish, in turn, will usually be less porous than a flamed finish.

3. What kind of abuse is the stone or tile likely to receive? A granite countertop or floor located in a heavily-used kitchen will require an oil-repelling penetrating sealer, while a polished marble foyer that is rarely walked on may only need an application of a water repellant penetrating sealer.

4. Does it shine? If the stone or tile has a dull finish, the application of a coating will add to its luster. One word of caution, however: if a coating is used to add shine, the maintenance required to maintain that shine will be higher than for the uncoated stone or tile.

Once you've considered all four of these factors, then, and only then, can you determine which sealer to buy. When choosing between a coating or a penetrating sealer, you must also bear in mind that coatings are generally less expensive than penetrating sealers, provide a sacrificial coating and can be applied below-grade. Many coatings will lend added-slip resistance and add luster and shine to the tile or stone.

On the other hand, coatings will scuff, scratch and mar very easily. Traffic patterns will be more noticeable. They can build up, causing an unsightly appearance; they may turn yellow and darken, and will require frequent reapplication. Most coatings will also have to be stripped and reapplied. Stripping chemicals can be dangerous to work with, and may cause damage to certain marbles and granites. Depending on the type of coating, they can also block the pores of some types of stone, causing a condition known as *spalling*. Spalling can be recognized by the development of pits and flaking of the stone surface. Coatings can rarely be used outdoors, since they are affected by ultraviolet (UV) light.

Penetrating sealers are generally more expensive than coatings, do not require frequent application, and can be used both indoors and out. However, many penetrating sealers are solvent-based, producing noxious odors and flammable vapors which can be extremely dangerous. The main advantage of penetrating sealers is that they generally do not change the appearance of the stone or tile and are very easy to maintain.

Marble & Tile

When choosing a sealer, it is always advisable to consult with your supplier, manufacturer or stone or tile specialist.

CHAPTER 9

Help For Architects & Designers

Help For Architects and Designers

"Will the tile and stone specified hold up under heavy traffic?"
"Is black marble softer than white?"
"Which finishes are more slip-resistant?"
"Can I mix stone and wood?"
"What installation specification are best?"
"Should I use epoxy or standard grout?"
"Should the tile be thin-set or mud-set?"
"How will the tile wear over time?"
"Who do I call for help and information?

These are just a few of the questions heard most often from architects and designers who design with tile and stone. The answers are not simple; a number of important factors must be considered before choosing tile or stone. Many installations have failed due to incorrect material selection. Some designs can also make maintenance impossible. It is extremely important that the professional designer ask the proper questions before specifying stone and tile. This chapter offers a resource guide to the organizations, associations and standards available for the selection and installation of tile and stone.

MARBLE AND STONE

Several stone and tile associations exist in the United States that can provide design manuals and recommendations for the installation and selection of stone, tiles and slabs. These organizations can also direct you to experts you can contact for help.

The Marble Institute of America (MIA)
30 Eden Alley, Suite 301
Columbus, OH 43215
614-228-6194
Fax 614-228-7434
E-mail: Rockoffice@aol.com

Publications:

• *Dimension Stones of the World, Volume I.* A resource, published in three-ring binder format, containing color plates of 312 dimension stones along with ASTM test values.

• *Dimension Stone Design Manual IV.* Specifications, technical information, typical details and a glossary of stone industry terms.

Technical Help: Consultation services

Marble & Tile

Allied Stone Industries
P.O. Box 360747
Columbus, OH 43236-0747
614-228-5489
614-224-5264 Fax

American Society for Testing & Materials (ASTM)
Committee C-18 on Dimension Stone
1916 Race St
Philadelphia, PA 19103
215-299-5400
215-299-2630 Fax

From the work of 132 technical standards-writing committees, ASTM publishes standard specifications, tests, practices, guides and definitions for materials, products, systems and services.

ASTM also publishes books containing reports on state of the art testing techniques and their possible applications.

American Society of Interior Designers (ASID)
608 Massachusetts Ave. NE
Washington, DC 20002
202-546-3480
202-546-3240 Fax

Marble & Tile

Association of Specialists in Cleaning & Restoration (ASCR)
10830 Annapolis Junction Rd.
Suite 312
Annapolis Junction, MD 20701-1120
301-604-4411
301-604-4713 Fax

Building Stone Institute
P.O. Box 507
24 Yerks Road
Purdys, NY 10578
914-232-5725
914-232-5259 Fax

Construction Specifications Institute (CSI)
601 Madison Street
Alexandria, VA 22314-1791
703-684-0300
703-684-0465 Fax

Indiana Limestone Institute of America
Stone City Bank Bldg.
Suite 400
Bedford, IN 47421
812-275-4426

The Indiana Limestone Institute was founded in 1928, 101 years after the opening of the first Indiana limestone quarry in Stinesville, IN.

ILI serves the construction sector, the architectural profession and the limestone industry as a coordinating agency for the dissemination of accurate, unbiased information on Indiana limestone standards, recommended practices, grades, colors, finishes, and all technical data required for specifying, detailing, fabricating, and erecting Indiana limestone.

The ILI publishes handbooks for both architects and contractors, as well as other publications which detail the proper methods for using Indiana limestone.

International Masonry Institute (IMI)
823 15th Street NW
Washington, DC 20005
202-783-3908
202-393-0219 Fax

International Society of Interior Designers (ISID)
1933 So Broadway
Suite 138
Los Angeles, CA 90007
213-744-1313
213-744-1252 Fax

Marble & Tile

Masonry Institute of America
2550 Beverly Blvd
Los Angeles, CA 90057-1085
213-388-0472
213-389-7514 Fax

Materials & Methods Standards Association (MMSA)
P.O. Box 350
Grand Haven, MI 49417
616-842-7844
616-842-1547 Fax

National Kitchen & Bath Association (NKBA)
687 Willow Street
Hackettstown, NJ 07840
908-852-0033
908-852-1695 Fax

National Stone Association
1415 Elliott Place NW
Washington, DC 20007
202-342-1100
202-342-0702 Fax

National Tile Contractors Association (NTCA)
P.O. Box 13629
Jackson, MS 39236
601-939-2071
601-932-6117 Fax

Marble & Tile

The National Training Center for Stone & Masonry Trades(NTC)
1413 Haven Drive
Orlando, Florida 32803
407-657-7878
Fax 407-898-0370
E-mail: Fhueston@aol.com
http://www.webcreations.com/marble

Provides training, seminars, publications and videos. Publishes *The Stone & Tile Report* newsletter.

Sealant, Waterproofing & Restoration Institute (SWRI)
3101 Broadway, Suite 585
Kansas City, MO 64111
816-561-8230
816-561-7765 Fax

Terrazzo, Tile and Marble Association of Canada (TTMAC)
30 Capstan Gate Unit 5
Concord, Ontario
Canada L4K 3E8
905-660-9640
905-660-5706 Fax

Tile Contractors Association of America (TCAA)
11501 Georgia Ave. #203
Wheaton, MD 20902
703-949-5995
703-949-8378 Fax

Tile Council of America (TCA)
P.O. Box 1787
Clemson, SC 29633
803-646-8453
803-646-2821 Fax

The TCA mission is to promote "Made-in-the-USA" ceramic tile and its proper installation. The Association provides technical assistance to architects and designers on the selection of tile, materials and installation methods. They are an independent testing laboratory for products of the industry. TCA publishes the highly regarded, industry-recognized *Handbook for Ceramic Tile Installation,* as well as the ANSI A108, A118, A136 and A137.1 specifications.

Several associations and organizations have developed standards, guide books and technical manuals for the installation and selection of tile and stone.

CHAPTER 10

Questions and Answers

QUESTIONS AND ANSWERS

About marble:

1. We live in San Francisco and are refurbishing our town house. We would like to use marble on our kitchen floor, but are concerned it may crack during an earthquake. Is there any way to install it that would minimize the cracking?

There are several *fracture membranes* that can be installed to help minimize cracking. These membranes are rubber- or plastic-based and are designed to absorb some shock; however, you will not totally eliminate cracking if the tremor is severe. Consult with your stone supplier or setting mechanic for further information and recommendations.

2. I have a black marble floor that I just cleaned with a wax stripper. As it dried, it turned white. I have tried rinsing it and nothing seems to work. Is it ruined, or can it be fixed?

It is probably not ruined, but removing this whitish haze can prove difficult. If the black marble is an agglomerate, chances are the alkaline stripper has attacked the polyester resin that binds it together. If so, the surface will have to be honed and polished. If the white haze is nothing more than a white film, a good wet-mopping using a stone soap will probably correct the problem.

3. My new marble floor has a white powder on it. It cleans up with water and soap, but keeps coming back. Is there anything I can use to clean it off permanently?

This a common condition, especially if the floor is new. When the marble is installed, it is set in a mortar base that is wet; as the moisture leaves the mortar, it carries with it dissolved salts which are deposited on the surface of the stone. This condition, called efflorescence, is what you are seeing as a white powder. Unfortunately, every time you clean the floor with water you are only dissolving more salt in the stone, and when the water dries, the salts return. To correct this problem, do not mop the floor with water. Either vacuum the powder or buff it off with a dry pad. The salts will eventually work themselves completely out of the stone. If you must wet-mop the floor, use a stone soap with cold water and wring the mop out tight, making sure to avoid puddling on the floor.

4. What type of dust mop can I use on my marble floor? I was told to stay away from treated dust mops.

The advice you received was correct. Most treated dust mops are treated with an oil-based chemical. Certain stones can absorb this oil, causing staining. Purchase only untreated dust mops.

5. What type of mop should I use on my tile floor? Sponge or string?

A string mop is preferable to a sponge mop. A sponge mop has a very small surface area, and can easily pick up sand and debris which will scratch the floor. It has been shown that after several swipes with a sponge mop, you are only spreading the dirt around the floor. Most janitorial supply houses sell good-quality string mops. I recommend a cotton-blend mop with sewn ends. The sewn ends allow the mop to be machine-washed. Be sure to use this mop only on your stone floor.

6. How do I remove the soap scum off my marble and tile shower wall?

Several products are available at the grocery store for removing soap scum. But be careful, as most soap scum removers contain acids which should not be used on marble surfaces. These cleaners work well on ceramic tile surfaces and are safe to use on them, but for marble and other stones I recommend contacting a stone care supplier (see "Resource Directory") and purchasing a product specified for marble.

7. I have a white marble vanity top in my master bathroom. It tends to develop white dull spots. What can I do to remove them?

These white dull spots are probably etch marks caused by cosmetics, shaving lotion, etc. To remove them, the vanity will have to polished, and, depending on the severity of the etch, possibly honed. I would recommend calling a stone professional, since honing and polishing are difficult procedures. Once the vanity top is restored, I would suggest sealing it with a good stone sealer to help prevent further etching.

8. Several weeks ago, our hot water heater developed a leak and flooded our kitchen floor. The floor is a white statuary marble, and it is starting to turn yellow. It is ten years old and we never had this problem until the flood. What's happening, and can it be fixed?

The problem you describe is very common with certain white marbles. Many marbles contain naturally occurring iron as part of their mineral make-up. When the stone is exposed to water for any length of time, its iron content begins to oxidize and turn to rust. The yellowing you see is the iron in the stone rusting. Under most circumstances, this oxidation will continue, and the floor will turn a steadily darker yellow and eventually a deep rich brown. The only solution is to replace the floor.

9. How do I clean the mildew off my marble shower wall with out dulling the shine? I have used soap-scum removers in the past, but they only dull the marble.

There are several products on the market that will remove mildew from tile, but they will also etch any polished marble surface. I suggest the following technique: Dilute one cup of household bleach in one-half gallon of cold water; add several drops of liquid dish soap. Wipe this solution on the mildew and let it stand for fifteen minutes, then rinse with plenty of cold water. If this does not remove the mildew, increase the standing time and repeat.

10. Is it safe to use pine-type cleaners on my marble floor? How about a ceramic tile floor?

Pine-type cleaners are nor recommended for polished stone surfaces . They contain harsh chemicals that can etch the stone. However, they work well, and are safe to use, on ceramic tile.

11. We installed a stone floor in my son's nursery. Because of his allergies, we have been told to mop-clean the floor with a disinfectant. What can we use that will disinfect without ruining the high polish on the marble?

I suggest you contact a local janitorial supply house and say that you need a disinfectant with a neutral pH. There are several dozen on the market that will do an excellent job.

12. The shine on our travertine marble floor has not looked even since it was applied. Many areas have dull spots. The installer told us this was characteristic of the stone, but I've noticed that a local hotel has the same travertine, and it looks shiny with no spots. What process could they have used, and is there hope for my floor?

Your installer is correct. Most travertine contains many holes that are filled with a cement filler. The cement filler does not take a shine when polished—hence the dull spots. The floor you observed at the hotel has probably been coated with a wax or made uniformly shiny through some other coating process. I would not recommend this for your home. A hotel spends every evening polishing and or waxing the travertine floor to maintain its shine; unless you would like to do the same, I would advise against it.

13. My marble vanity top has developed several chips along its edges. Can these be repaired?

Possibly. Have your local stone professional take a look at it. If the edge is square, he (or she) may suggest rounding the edge to prevent further chipping.

14. Ever since my ceramic tile floor was installed, when I walk across it wearing heels, I can hear a hollow sound under several tiles. Is this normal, or should I be concerned?

You should have a professional installer check the installation for soundness. It may have not been installed properly. The hollow sounding tiles is an indication that areas under the tile are not in contact with the setting bed. These hollow areas have the potential to cause cracking.

15. Is it safe to use a vegetable oil-based cleaner on my marble floor?

Most of the stone soap products sold by stone care suppliers contain vegetable oil-based soap, and they are safe for marble surfaces. Remember not to overuse them, and to follow the manufacturer's instructions, since too much soap can produce streaking and a build-up which could make the floor slippery.

16. On a bright, sunny day, I can see light circular swirls in my marble floor. What causes these swirls, and can they be removed?

If your floor has ever been polished, these swirls are the result of the polishing method that was employed. If steel wool or an abrasive pad was used, it created light swirl scratches in the floor. These can be removed by honing the floor and repolishing it. Call your stone professional.

17. The white veins in the black marble

floor in my bathroom appear to be crumbling and disintegrating. What's causing this, and can it be repaired?

Black marble, especially Negro Marquina, is very susceptible to spalling. Spalling is a condition in which the stone flakes off and falls apart. Several factors can cause spalling: excess moisture, choice of setting material and method, the polishing process, etc. I suggest calling in a professional to inspect the floor. If the spalling is not severe, it may be possible to repair it by filling the missing veins with a polyester fill.

18. After I mop my tile floor, there are streaks across it when it dries. I have tried several different cleaners, yet I always end up with these streaks. Any suggestions?

The streaking may be caused by any one of a number of problems. The mop you are using may be dirty; buy a new, clean string mop and make a point of only using it for the marble floor. Wring the mop out several times as you clean the floor. Try switching to a stone soap instead of the cleaners you are using. Be sure to follow the directions; too much cleaner will cause streaking. Use only cold water, since hot water will also cause the floor to streak.

19. Is it safe to use a store-bought cleaner/wax combination on my marble floor?

I would advise against it. Although these are excellent products for vinyl floors, most of them have a tendency to yellow and darken marble. If you must wax your floor, I would suggest that you contact a stone care supplier and purchase a product designed specifically for marble.

20. How do I clean the grout on my marble floor without dulling the marble? All grout cleaners seem to contain acid.

Several excellent grout cleaners are available from your stone care supplier that will not dull polished marble. Try a mild solution of bleach (see question #9 above); use a soft tooth brush to apply the solution to the grout joints and scrub lightly.

21. Is it possible to polish my marble floor myself? Can I rent or purchase the necessary machine and supplies?

Unless you are trained in the art of stone restoration, I would not recommend polishing your own marble. Too many problems can occur, and you might end up turning a simple polishing job onto a costly restoration. Call a reputable stone supplier.

22. Is it safe to use household bleach on my white marble floor?

I would not recommend using bleach on any marble floor as a daily cleaner. Household bleach can cause staining on certain types of marble. Use only a good stone soap or neutral cleaner.

23. I just bought a house that has a beautiful marble floor. The floor appears to have had many coats of wax applied to it. Since I have no experience with marble, please tell me what I need to do to strip off the old wax. Do I need to wax it again, and with what?

Many wax strippers are commercially available. Any janitorial or stone supply house can recommend several. Always test the stripper in a small area to make sure there are no adverse effects. Chances are that when all the wax is removed, the marble will appear dull. At this point there are two options. The marble can be re-coated with additional wax; if you choose to do this, be sure to purchase wax specifically designed for stone. The other option is to have the floor refinished and repolished. Repolishing option is generally is the best alternative, since the shine will match that of a new floor and will be easier to maintain.

24. I own a white marble statuary bust that is yellow and dingy-looking. What can I use to brighten it up?

If the sculpture is very valuable, I would not advise you to attempt to clean it yourself. Old marble sculptures can be made out of marbles that contain iron. Over time, the iron becomes oxidized and turns yellow. If the yellowing you are observing is the result of iron staining, it cannot be removed. If the yellow is only surface dirt, try cleaning the bust with a solution of stone soap or neutral cleaner. A soft toothbrush can be used to get into all the nooks and crannies.

25. The marble under the carpet in my front foyer has yellowed. What can I use to remove the yellowing?

The yellowing is the result of moisture becoming trapped under the carpet. The carpet's jute backing or dyes from its rubber backing have bled into the stone. To remove this yellowing, try the following steps:

- Clean the area with a solution of an

alkaline stone cleaner. Be sure to rinse thoroughly. If this fails, a poultice may have to be applied.

• Apply a poultice of diatomaceous earth and a solvent such as mineral spirits. For detailed instructions, see the poultice section of this book.

26. Can I use glass cleaner to clean my marble shower walls?

No, I would not recommend using window cleaner on a marble shower wall. Some window cleaners contain chemicals which could damage the marble. Use only a good stone soap or neutral cleaner.

27. How do I clean around the fixtures on my marble vanity top?

This has always presented a problem. If the marble adjacent to the fixtures is severely pitted, I would recommend hiring a plumber to remove the fixtures and a refinishing contractor to refinish the vanity. Once the marble has been refinished, use a good stone soap or neutral cleaner to clean around the fixtures. Avoid build-up of soap by cleaning behind and around the fixtures daily with the stone soap. If soap scum accumulates anyway, a marble supplier can suggest several products for removing it. Make sure to read the labels carefully, and to avoid acid products, since these could tarnish the metal fixtures.

28. My dog is not yet housebroken, and is constantly urinating on my marble floor. I've found that I can clean off the urine without staining, but a dull area is left afterwards. Also, how can I eliminate the odor?

Urine contains acid which will etch the marble surface. To restore the finish, the marble will have to re-polished using a marble powder polish.
Fortunately, pet shops and supermarkets carry several products which use enzymes to destroy the odor of urine. Follow the manufacturer's directions carefully, and apply the odor destroyer before re-polishing.

29. My bathroom marble floor is very slippery when I get out of the shower. Is there any treatment I can put on the floor to keep it from getting slippery?

Although there are several coatings that will provide added slip resistance, there is, unfortunately, a trade-off. Coating the stone will increase maintenance requirements, but if safety is an issue, it's worth considering. Contact a stone maintenance supplier or janitorial supply for recommendations.

30. There is a wide, dull spot on my marble foyer where the front door opens and closes. How can this be fixed?

The marble has probably been scratched by the underside of the door. First, trim the door bottom so that it does not scrape on the marble. Then have the dull spot refinished and polished. I would suggest calling a professional stone restoration company.

31. I've noticed little white specks on my white marble floor. They seem to be within the marble and you can't feel them. What are these, and how can they be removed?

These white specks are called stun marks, and cannot be removed. They are the result of sharp objects hitting the marble causing an explosion of the crystals in the marble. High heels are a common cause of these marks, so avoid wearing them when walking on the marble.

32. We used to have a large fichus tree on our marble floor. When it died, we removed it and found a brown stain where the plant had sat. What should we use to remove this stain?

The brown stain is the result of minerals from the soil of the fichus tree. This type of stain is difficult to remove and will require a poultice. Refer to the Stain Removal chart in Chapter eight and apply the poultice as instructed.

33. I just applied a silicon sealer to my marble floor, and it is sticky; I can't seem to clean it up. What can I do?

The sticky substance is silicone residue, which can be difficult to remove. First, try cleaning the floor with mineral spirits and clean cloths. If this does not remove the silicone, then solvents may be necessary. Suggested solvents may be acetone, toluene or methylene chloride. Be careful; these solvents are very hazardous to work with, and can be explosive.

34. My four-year-old got hold of a black permanent marker and drew all over my white Thassos marble floor. How do I clean it off?

Depending on how long the marker has remained on the floor it may require a poultice. Mix methylene chloride and poultice powder and apply. Refer to chapter eight on stain removal.

35. How can I clean the black soot off my marble fireplace hearth?

Several products are designed specifically for cleaning soot and carbon build-up from marble. Contact your marble supply house. Be sure the product is not an acid; apply it per the manufacturer's directions and allow it to sit on the marble for 10–20 minutes. Agitate lightly with a soft-bristle brush and thoroughly rinse it from the surface. Repeat this procedure if the first application does not clean all the soot from the marble.

36. What type of floor mats do you recommend for use on my marble floor? I was told to stay away from rubber-backed mats.

You received sound advice. Rubber-backed mats, as well as some jute-backed mats, can stain your marble floor. I would advise using only cloth-backed floor mats. Be sure to take them up at least once week to avoid moisture build-up.

37. The green marble floor in our bathroom is acquiring a specked white appearance that won't clean off. What can I use to clean it?

First, you need to determine whether or not the white specks you describe are on the surface of the marble. If you can't clean them off, chances are they lie below the surface of the marble. If this is the case, it is possible that there are dissolved salts in the marble—and should this condition persist, the marble will become pitted. I would call a stone restoration specialist who can diagnose the problem. He will probably recommend thoroughly drying the marble and then applying a silicone impregnator.

38. We just had our floor professional polished and now it is turning yellow. What can we do?

This is a frequent problem with professional polishers who use several different polishing methods. Have the professional come back and correct the problem. If he used the recrystallization process, the floor may have contained too much moisture. He will have to re-polish the floor with polishing powder to remove the yellowing, and if he plans to use the recrystallization process again, the floor will have to be allowed to dry for several days before reapplication.

39. My husband spilled a glass of iced tea over a newspaper that was on top of our marble dining room table. The newspaper ink has stained the table. How can I remove this stain?

The newspaper ink has probably soaked into the marble, and may require a poultice to remove, but first try cleaning the area with a white cloth and some acetone. Be careful with the acetone, since it can be flammable, and keep it away from any painted surfaces. Blot the stain with the acetone and rag; if the stain does not come out, prepare a poultice of diatomaceous earth and mineral spirits. Refer to the chapter on stains in this book for detailed instructions.

40. How do I remove pencil marks from a white marble stair?

Pencil marks can frequently be removed simply by erasing them with a pencil eraser. If the eraser trick does not work, then a poultice of denatured alcohol and a poultice powder will need to be applied.

41. Our green marble floor has never looked very shiny in comparison with the black marble in our bathroom. Can it somehow be given a very deep shine like the other marble?

It depends on what kind of green marble you have. I would guess that the green marble in your bathroom is an Asian green. If this is the case, you will never get a deep shine on this marble. These stones characteristically have a clouded shine. If, on the other hand, you have a Vermont green or something similar, it may need to be professionally polished.

42. How do I prevent my dining room chairs from scratching my marble floor?

The chair legs need to have plastic or cloth felt protectors attached to them. These protectors are small square or round tabs that are designed to prevent scratching. They are available at most hardware and home supply centers.

43. A lemon slice sat on our marble bar top overnight and left a white mark. How can we remove this mark?

The acid in the lemon has etched the surface of the marble. The only option is to re-polish or hone the marble. There are several do-it-yourself kits on the market which can improve the etch, but I would suggest the services of a stone specialist if you want it done right.

44. The veins on our white marble shower are turning yellow. How can we remove the discoloration?

There are two possible reasons the veins in your marble are changing color. First, white marble characteristically contains iron; it is possible the iron is oxidizing and turning to rust. If this is the case, it may be impossible to remove. Second, have your water checked for iron. If your water contains iron it may be being deposited in the veins of the marble. If this is the case, it may be possible to remove the iron using some of the commercially available iron removers. But if you decide to use one of these formulas, proceed with caution; many contain strong acids which can damage the marble.

45. Where can I buy marble care products? The local grocery stores don't seem to carry any.

Marble and stone care products can be purchased from a stone supplier or janitorial supply company. If you have trouble locating any, call a few marble and stone distributors and ask them where these products can be purchased.

46. How often should I have my marble floors professionally polished?

That's a difficult question, since it depends on the type of marble you have, how hard or soft it is, how much traffic the floor receives, and many other factors. On the average, most marble floors in a home need to be polished about once every one or two years.

47. Is it all right to use a solution of vinegar and water on my marble floors? What about my ceramic tile?

Vinegar and water make a great cleaner for ceramic tile, but should never be used to clean marble. Do not use vinegar on marble! Vinegar contains acetic acid, which will dull the finish.

48. Our travertine floor is developing black dull spots that won't come out. What can we do?

Travertine marble typically contains soft fillers that have a tendency to collect dirt. The black spots are probably dirt, and can be cleaned with heavy-duty cleaners. Refer to this book's section on heavy-duty cleaning; if the problem is severe, re-honing and polishing may be necessary.

49. I just had my concrete pool deck acid-washed. One of the workers walked through the house and stepped on my marble floor, leaving footprints which won't come off. What can I use to remove them?

The acid has etched your marble, which will now have to be polished or (possibly) honed. This is a task that requires the services of a professional; break out the Yellow Pages and start dialing.

50. We just installed a new green marble foyer. The problem we are having is that the tiles are curling up on the corners. Why is this occurring, and can it be corrected?

Curling in green marble can be a serious problem when the marble is set in a wet mud or thin-set. The Marble Institute of

America recommends that green marble be set with an epoxy-based compound. Check with the installer and see what has been used to set the marble. If mud or thin-set was used, the marble will have to be removed and reset in epoxy.

51. A number of marble professionals have given me a bid for polishing my marble floor. Several of them have mentioned the word recrystallization. What is this process, and is it safe for my marble?

There are several processes for the polishing of marble, and recrystallization is one of them. Refer to my chapter on polishing for a complete description of this process. The safety of the process is as safe as the applicator. Be sure the marble contractors is familiar with this process and ask for references where he has used it before.

52. How can I find a reputable marble polisher? Is there an organization I can call for a reference?

There are several associations that can refer a contractor in your area. I have listed several below. Always ask for references from previous customers to make sure you are using a reputable contractor.

Marble Institute of America
30 Eden Alley #201
Columbus, OH 43215
614-228-6194

Building Stone Institute
PO Box 507
Purdys, NY 10578
914-232-5725

53. My marble contractor tells me my floor may have "v-edges" when he gets done polishing it. What is a v-edge?

V-edge is a term used by professional stone refinishers to describe the shiny edge that may remain after polishing. V-edges result from uneven tiles. When the tile is polished, the polishing pad can not reach the lower edge of the uneven tile, and that lower edge may have a different shine then the remaining tile. It is common for this lower edge to be at the corner of the tile forming a "v"—hence the term.

54. What is the difference between marble and cultured marble and Corian™?

Corian™ and cultured marble are man-made materials consisting mostly of plastic resins. Marble is a natural substance extracted from the earth. For a complete description, refer to this book's section on stone and tile types.

55. Can Corian™ and cultured marble be treated like real marble?

In many ways the care and restoration of these materials and natural marble are the same. For daily care and cleaning, treat them as you would marble. But for restoration, consult a professional who is familiar with them.

56. I have a large marble insert surrounded by a pine wood floor. The marble needs to be refinished, but I am concerned that any water will ruin the wood floor. Do you have any suggestions?

This is a difficult problem, since water is usually required to refinish marble, and water will warp the wood. But there are several options. First, consider refinishing the marble using dry abrasives, such as sanding screens; these will create dust, but will not warp the wood. If the use of water is absolutely necessary, it may be possible to dam the wood with plastic and seal the plastic dam with silicone caulking. Whichever option you choose be sure to hire a reputable contractor to do it.

57. How often do I need to reseal my marble shower wall?

This depends on several factors, including how often the shower is used and what type of sealer was originally applied. After identifying the product, refer to the manufacturer's recommendations. If water is no longer beading on the shower walls, sealer reapplication would be advisable.

58. We have a marble bathroom floor and keep getting spots around the toilet. Is there any sealer we can place on the marble to prevent this spotting?

The spotting can be caused by dripping toilet bowl cleaners, which can etch the marble surface. As of this writing there are no sealers that will prevent acid etching. I would suggest purchasing a non-acidic bathroom bowl cleaner, available at most janitorial supply stores.

59. Is there a way to remove cigarette burns from my marble vanity top?

Cigarette burns can only be removed by honing and repolishing the marble. Purchase one of the do-it yourself repair kits, or call in a professional.

60. Is there a non-acid toilet bowl cleaner I can use? My marble floors are being ruined by acid cleaners.

Most janitorial supply stores carry non-acid bowl cleaners.

61. How do I remove etch marks from a marble shower wall? The maid used an acid cleaner on them and it left some deep marks.

Since the acid has etched the marble, the etching will require honing and polishing to remove. Call in a professional stone specialist, since this work can be quite difficult.

62. We live in a very sandy area, and the sand is severely scratching our marble floor. How can we keep the sand from getting on the floor?

Unfortunately, there is nothing you can put on the floor that will prevent it from scratching. You may find it useful to place several walk-off mats, both inside and outside the entrances, to help trap most of the sand. Dust-mopping the floor as often as possible should also help.

63. We installed a marble shower wall, and the contractor used an acid to clean it. We had it professionally repolished, but it keeps getting etched after we take a shower. What's causing the etching and how do we correct it?

I will assume you are not cleaning it with acid cleaners. There is a possibility that there is some residual acid in the grout or the pores of the stone that bleeds out when you take a shower. Try rinsing the shower wall with a solution of baking soda and water. Prepare approximately one cup of baking soda to one gallon of water, and flood the wall several times with this solution. If there is any residual acid, this solution should neutralize it.

64. Is it OK to use a vacuum cleaner on my marble floors?

Yes and no. A vacuum cleaner can be an excellent tool for eliminating dust and dirt, but you must be sure that it will not scratch the marble. Check the bottom of the vac and make sure there will be no metal rubbing on the stone as you move it back and forth over the floor. Check the beater brush to make sure it will not scratch the marble. Always use the soft brush when using hose attachments and never drag the vac across the floor.

65. My husband is a heavy smoker, and his bad habit has stained my white marble walls yellow. How can I remove these stains?

Smoke stains can be difficult to remove, and may require the use of a poultice. There are several gel-type cleaners that will remove smoke stains; ask your stone supplier for a recommendation. The gel cleaners are brushed onto the surface and allowed to sit for several minutes before being rinsed away. This sounds like a job that should be performed by a professional.

66. Several marble tiles next to our toilet are stained with urine. Is there any way to remove them? The marble has been stained for several years.

The likelihood of removing these stains is about the same as that of winning the lottery. I would suggest replacing them.

About granite:

1. Every time it rains, a blotchy light-and-dark stain appears on my granite sidewalk. If it stays dry for several weeks, it goes away. What is happening, and can I put something on it so that it doesn't occur again?

Most granite is very absorbent, and will soak up a large quantity of water. You are seeing moisture in the granite, which can take anywhere from ,several days to several months to dry out. Once the granite dries apply a silicone-based penetrating sealer. This should keep the water from passing through the surface of the granite.

2. How do I remove chewing gum from my flamed granite sidewalk?

You can remove chewing gum by freezing it. Place ice cubes, or, if available, dry ice on top of the gum. The gum will freeze hard, and can easily be popped off the granite.

3. Someone told me I could use car wax on my granite vanity top. Is this true? If not, what should I use to protect it and keep the shine?

Most of the marble and stone polishes on the market contain the same type of wax found in everyday car wax. However, I would be cautious about applying car wax to a stone countertop; the newer car waxes contain colorants which could stain the stone. I would suggest purchasing a stone furniture polish made specifically for stone.

4. My shower floor is made of a flamed granite. A hard white deposit is starting to form between the grout lines. What is this, and what do I use to clean it?

The hard white deposit is a build-up of salts. These salts—known in the tile industry as "lime putty"—are coming up from the setting bed, and can be very difficult to remove. To remove lime putty, scrape away as much of it as possible using a putty knife or razor blade. Next, use a solution of muriatic (hydrochloric) acid in a dilution of 1:1. This should dissolve the lime putty . Be very careful, and make sure there is adequate ventilation. Hydrochloric acid emits caustic fumes which are very dangerous. Wear an approved acid fume respirator and use an exhaust fan. Also, protect any metal fixtures, since the acid fumes will tarnish metal surfaces. This is a job for which you should definitely consider hiring a professional. When all the lime putty has been removed, and the tile is completely dry, apply a good-quality silicon sealer to help prevent reoccurrence of the putty.

5. My granite vanity top has gold/brass fixtures. What can I use to clean the granite without damaging the fixtures?

For daily and light cleaning, use a good stone soap, which will not harm the gold or brass fixtures. For heavier-duty cleaning, remove the fixtures or protect them with masking tape. Be careful when using acidic cleaners, since acids will tarnish the fixtures.

6. The silicone caulking around my granite shower wall seems to be staining the granite. Is there a way to remove this stain?

Staining from silicone can be very difficult to remove, but not impossible. First remove all the silicone caulking; removal of the stain will require the application of a poultice and a strong solvent such as methylene chloride. Several applications may be necessary. For further details, refer to the stain removal section in chapter eight.

7. The granite in my master bath shower is pitting and feels very rough. What is causing this, and can it be fixed?

There are several possible causes for the pitting. Improper cleaning chemicals and salt deposits can contribute to this condition. I would suggest having a reputable stone restoration firm examine the shower and suggest a remedy. Depending on the severity of the pitting, replacement may be necessary.

8. My granite floor appears to be developing cracks beneath its surface. These cracks do not go all the way through the tile, but can be seen only in a bright light and only at certain angles. It almost looks as if they're cracking from the bottom, and the crack hasn't quite reached the surface. Should I be concerned?

The condition you are describing is called compression cracking. It is caused by movement of the substrate to which the granite is attached. If the granite tile is set over a plywood floor supported by joists, movement in the floor may be responsible. You should be concerned but not alarmed. Compression cracking may lead to cracking which would require replacement of the cracked tiles. I would keep a close eye on the cracks, and if they develop further, would recommend consulting a stone specialist.

9. How can I remove fingernail polish from my granite vanity?

The procedure depends on how long the nail polish has been allowed to set. First, try removing it with acetone and a clean rag. If it has dried completely, it may be necessary to scrape the polish with a sharp razor blade, followed by cleaning with acetone.

10. When it rains, I notice a dark stain running around the perimeter of each granite tile on my back porch. This outline doesn't seem to ever go away. What's causing this and can it be cured?

This shadow or halo is a sign of moisture in the granite. When the stone gets wet, it turns dark; the granite then dries from the center out, leaving a moisture shadow which has the appearance of a stain. To eliminate this shadow, the granite will have to be dried completely, followed by the application of a good silicone impregnator.

11. How should a newly-installed flamed granite floor be cleaned for the first time?

Flamed granite has a rough texture which will tend to hold dust and dirt. Prepare a solution of stone soap or neutral cleaner and warm water. Flood the floor with this solution and either agitate with a scrub brush or machine-scrub. Pick up the solution with a wet-vac and repeat until clean. After the floor has had time to dry, apply a good-quality silicone sealer.

12. The granite panels above our fireplace have developed several dark circles near the corners of each panel. The granite is only three months old. Should I call the contractor and have them replaced?

I would call the contractor, but it may not be necessary to replace the panels. These dark circles may be the result of moisture bleeding from the setting mortar. It is common for panels to be installed by spotting the backs of the panels with setting material. It may be that this setting material has not thoroughly dried, and you are seeing the moisture bleeding through. To be sure, have the contractor take moisture readings. Moisture in granite can take months to dry out completely.

On slate:

1. What is the best way to clean a gray slate floor? It streaks every time I mop it, and the mop seems to shed on the floor.

It sounds as though your slate floor has a rough finish which catches the fibers of the mop as it is dragged across the floor. First, you need to deep-clean the floor to remove all dirt and grime. Next, apply a good-quality acrylic sealer. Several coats may be necessary. The acrylic sealer should make future cleaning much easier. Clean the floor with a neutral cleaner and re-apply the sealer as needed.

2. The slate steps on our front porch have developed a hard white deposit coming from the grout lines. How can we clean it off?

The white deposits are salts, which are bleeding through from the grout joints or from the mortar. An acid wash will be required to remove the salts. If the deposits are heavy, try scraping them off before washing the slate with acid. Several acid products are available for this purpose. Ask your local stone supplier.

3. What's the best way to remove a polyurethane finish that was placed on my slate floor?

Unfortunately, most polyurethane finishes require strong solvent-based chemicals for removal. Methylene-chloride strippers as well as a number of safe strippers are available, but be forewarned: removing polyurethane is a messy job. I know several professionals who refuse to do it!

On terrazzo:

1. My terrazzo floor has a yellow cast to it. How can I remove this?

First, you need to determine what is causing the yellow cast. If a carpet was at one time placed over the terrazzo, the jute or rubber backing may have bled color into the floor, turning it yellow. If the floor was waxed, the waxes will have yellowed over time. To remove the yellowing, try the following:

- If the yellowing is due to old wax, strip the floor with a floor machine equipped with a stripping pad using a good-quality floor wax stripper (available at most janitorial supply houses).

• If the yellowing is the result of carpet-back bleeding, it will be more difficult to remove. There are several chemicals on the market that are designed for removing yellowing due to carpet backs. These products are usually solvent-based and can be very difficult to work with. I would suggest calling a floor or stone restoration professional.

2. We have an old terrazzo floor that has been covered for years with wall-to-wall carpet. We want to remove the carpet, but it is held down by tack strips which have been nailed into the terrazzo. When I tried pulling up one strip, large chunks of the terrazzo came with it. Is there a way to remove these strips without damaging the terrazzo?

The tack strips can be removed by cutting the nails flush with the terrazzo. A hack saw blade can be wedged under the wooden strip and the nail can be cut. A faster method would be to use a diamond saw blade mounted on a right-angle grinder. The shaft of the nail will remain, but will go almost unnoticed.

3. Can cracks in terrazzo be repaired? How?

Yes, cracks can be repaired. Depending on the width of the crack, It can be filled with a polyester resin. If the crack is small, clean it thoroughly with acetone, making sure to remove all dirt; a dental pick may prove useful here. After the crack is clean, mix a polyester fill (available at marble supply houses) and work it into the crack with a razor blade or flat-blade trowel. For a small crack, I would suggest using a clear polyester. If you are uncomfortable with this procedure I would suggest calling in a professional.

4. How do you remove rust stains from a terrazzo floor?

Removal of rust on terrazzo will typically require both a powder poultice and a commercially available rust remover. Refer to the stain section in this book for detailed application procedures.

5. Our terrazzo floor has had many coats of wax applied to it over the years. The wax build-up on the terrazzo baseboards is very thick and difficult to remove. Do you have any suggestions?

Wax build-up on a vertical surface can be difficult to remove, since most strippers will run. Janitorial supply stores do offer spray-on baseboard wax removers that work very well; since they are applied as a foam, they will cling to vertical surfaces, allowing for easy removal. If you cannot find any of these formulas, a spray-on oven cleaner would be a good substitute.

6. The color on my red terrazzo floor has faded. Is there a way to bring the color back?

There are several techniques that can be used to bring it back, but first it is necessary to determine why the floor has faded. If the fading is due to excessive wear from foot traffic, resurfacing and polishing should return the color. If the fading is caused by sunlight, then the terrazzo can be enhanced with one of several sealers designed for this purpose. Consult a stone specialist for recommendations.

On ceramic tile:

1. Ever since my ceramic tile floor was installed, when I walk across it wearing heels, I can hear a hollow sound under several tiles. Is this normal, or should I be concerned?

You should have a professional installer check the installation for soundness. It may have not been installed properly. The hollow sounding tiles is an indication that areas under the tile are not in contact with the setting bed. These hollow areas have the potential to cause cracking.

2. Is it safe to use pine-type cleaners on ceramic tile floor?

As already noted in the section on marble, pine-type cleaners are effective and safe to use on ceramic tile.

3. My ceramic tile floor has developed several white streaks. No matter how I clean them or what chemical I use, they don't seem to go away. What can I use to clean them off?

These white streaks may be hard-water deposits, which can prove very difficult to remove. Try cleaning with a sulfamic acid-based tile-and-grout cleaner (available at your tile supply store).

4. My ceramic tile kitchen floor has several chipped tiles. Replacement tile is no longer available. Can these chips be fixed, and if so, how?

Chipping in ceramic tile can be difficult to repair. There are several repair kits now available that are composed of polyester resins. If these kits do not do the trick, call a professional tile restoration expert.

5. How do I keep the grout on my ceramic kitchen counters clean? It seems to be a never-ending battle.

Several grout cleaners are available at your home center, hardware store or janitorial supply store. These cleaners are acidic, so caution is advised. Once the grout is cleaned I would suggest applying a good grout sealer. You may also want to consider removing the existing grout and replacing it with one of an epoxy type. Epoxy grouts are stain resistant and easier to keep clean.

6. Is it all right to use a solution of vinegar and water on my ceramic tile?

Vinegar and water make a great cleaner for ceramic tile, although, as already noted, they should never be used to clean marble.

7. We just sealed our ceramic tile with a water-based silicone sealer. There is now a light, rainbow-like film on the surface of the tile. I have used every cleaner I could purchase to try to get it off, but it won't budge. Any suggestions?

The sealer has left a hard silicone residue on the tile's surface. To remove the excess silicone, the tile will have to be scrubbed with an abrasive cleaner. Several kitchen-type abrasive cleaners are available which will remove this film. If the area is large, a floor machine may be needed. Work the abrasive cleaner into the tile, adding just enough water to form a thick paste. Scrub with a circular motion and rinse the tile thoroughly with clean water. Several cleanings may be necessary to remove all of the film. If this procedure does not work, turn the job over to a professional, who will use more aggressive methods.

8. What can I use to seal the grout on my ceramic tile shower walls—and is this something I can do myself?

Many brands of grout sealer are available for this purpose; most are easy to apply, but make sure the grout is first cleaned thoroughly. Grout sealers are available at most hardware and home supply centers.

9. The grout on my ceramic tile floor is discolored. I tried cleaning it, but the colors are now different. Is there a way to color the grout?

There are several grout paints and dies on the market. But be aware that coloring grout can be a tedious task requiring great patience and a steady hand. It is also extremely important that the grout be thoroughly cleaned; otherwise the color will peel away. This is a difficult do-it-yourself project, and I would highly recommend consulting a professional.

10. Several years ago we had the grout on our ceramic tile floor colored. Now the color is peeling off, and it looks terrible. Can I have it recolored?

Yes, it is possible to recolor the grout, but it needs to be thoroughly cleaned before re-coloring. If possible, find out what brand and type of coloring was used, and re-apply the same brand or type.

11. Is there any way to prevent hard-water spots from forming on our ceramic tile shower walls?

The water spots are, of course, literally caused by the hard water . There are several silicone waxes that can be applied to the tile to help prevent them from forming. A vinegar-and-water mixture works well for removing these spots.

12. Help! My ceramic tile floor has lost its glaze. Is there a process available to restore the glaze, or do I have to replace the floor?

At this time there is no process that can fully restore the glaze to a ceramic tile floor. The glaze on ceramic tile is achieved by baking the tile at very high temperatures which cannot be achieved after installation. There are, however, a number of coatings and polishing processes that can improve the shine and thereby delay replacement of the tile. Contact your local tile store or tile contractor for further information.

13. We have a terrible problem with hard-water deposits in our ceramic tile shower. We were told that installing a water softener would solve this problem. What is your recommendation?

I definitely agree. A water softener will remove most of the minerals that cause hard-water deposits—and you will probably use less soap and shampoo, avoiding soap build-up as well.

On Mexican tile:

1. I have a white pickled Mexican tile floor that is peeling. Can it be re-pickled?

Pickled Mexican clay tiles are pickled at the factory. It is difficult to re-pickle the tile after it has been installed. This is a job for a professional, and it will be very expensive, since all the grout will have to masked to keep the pickling paint from running into it.

2. My Mexican tile floor turns a cloudy white whenever I spill water on it. What's causing this, and can it be fixed?

This cloudy-white condition is common on tiles that have been coated with an acrylic wax. Consider having the floor stripped and re-coated with a good-quality Mexican tile sealer.

3. The Mexican tile floor on my back porch is turning yellow. Help!

First, has the tile been sealed? There are several urethane coatings on the market that are used to coat Mexican tile. These urethanes are sensitive to ultraviolet (UV) light and will turn yellow. Strip the floor and apply a Mexican tile sealer. Warning: Urethane coatings can be extremely difficult to strip, requiring strong solvents for removal.

4. I can't seem to mop my Mexican tile floor! It sucks up water as quickly as I put it on. Any suggestions?

Again, has the tile been sealed? It sounds as if it hasn't been. Clean the floor thoroughly and apply several coats of a good-quality sealer. If the floor has been sealed, then it does not have enough sealer. Clean it and apply additional sealer.

5. How do I remove oil spots from my Mexican tile floor? Grease spattering from the stove is causing staining all over the floor.

Mexican tile is a clay terra cotta-type product, and is very absorbent. In order to remove the oil a poultice is necessary. Refer to the stain-removal section of this book and carry out the procedure it describes. If the stain is not removed after four or five attempts, I would suggest replacing the stained tile.

6. We have a Mexican tile deck all around our swimming pool. It is developing a whitish cast to it. Can this be cleaned and with what?

This white cast is very common with Mexican tile. Hard water and chlorine from your swimming pool are depositing a white salt on the surface. To remove this white haze, a mild acid washing is necessary. Use a mild acid such as sulfamic acid (available at most home centers or tile supply stores). Mix the acid solution and apply it to the tile. Scrub with a soft, brittle brush and rinse thoroughly. Be sure to follow label directions carefully. Acids can be difficult to work with, and can cause damage to other surfaces, such as metal. If in doubt, hire a professional to do the job.

7. The Mexican tile on our back porch is displaying a green growth at the corners of the porch. The porch is outdoors, but is covered. What is this growth and how do I clean it off?

The green growth you see is probably algae. Algae usually grows were there is abundant moisture. My guess would be that the tile stays wet after a rainstorm, allowing the algae to grow. To clean it, prepare a solution of bleach and water; about one cup of bleach to a gallon of water should do the trick. Scrub the tile with this solution and rinse thoroughly. To prevent the algae from returning, it is advisable to seal the tile. Visit your local janitorial supplier and ask for a sealer that can be used outside. Be careful on this point, as many sealers are not recommended for exterior applications.

On miscellaneous products:

1. What is the best way to apply silicone sealers? Can I mop it on?

There are several methods of sealer application. Most sealers can be sprayed on, mopped on or applied with a lamb's-wool applicator. I prefer the lamb's-wool applicator because it prevents streaking. Before choosing a method of application it would be advisable to consult the manufacturer of the product you plan to use.

2. Is there a product I can use to keep the soap scum off my shower walls?

There are many on the market. But if your shower walls are marble, you will have to use caution in choosing the proper cleaner, since most soap scum removers are acidic. Do not use an acidic cleaner on marble walls! If the tile is ceramic or granite, most soap-scum removers are safe to use. To prevent soap scum from developing, apply several coats of a good-quality silicone impregnator to the shower wall, making sure to cover the grout lines where soap scum tends to accumulate.

3. The caulking on my tub surround has turned black. I have tried all kinds of mildew cleaners, but nothing works. I hope you have a suggestion!

Mildew can be difficult to remove from silicone caulking because it grows into the caulking. I suggest removing the caulking with a putty knife and replacing it with new caulk.

4. How can I remove mildew from the grout on my shower walls? I've tried all the mildew removers, and they don't do a good job.

The mildew on your grout has penetrated deeply. To remove it, you will have to prepare a mixture of bleach and an abrasive cleaner like Soft Scrub. Apply the thick mixture to the grout with a paint brush and allow it to set overnight. Rinse the thick paste away and reapply if necessary. If this procedure doesn't work, the grout will have to be removed and replaced.

5. How do I remove the soap scum off my marble-and-tile shower wall?

Several products are available at the grocery store for removing soap scum. But be careful, as most soap scum removers contain acids which should not be used on marble surfaces. These cleaners work well on ceramic tile surfaces and are safe to use on them, but for marble and other stones I recommend contacting a stone care supplier (see "Resource Directory") and purchasing a product specified for marble.

And the inevitable:

1. We are remodeling our kitchen and want to put in new countertops. Which is better, marble or granite?

Both granite and marble make

excellent countertops, depending on the counter's intended use. Marble is sensitive to acids, which can make it difficult to keep polished in a kitchen. Granite, by contrast, is acid-resistant, and can make an excellent counter in a well-used kitchen; on the other hand, it can be very porous, and will thus stain with oils. Be sure to seal it with a good quality silicone sealer.

2. Can marble and granite be mixed on a floor? I have seen several patterns that employ marble/granite combinations.

Marble and granite can certainly be mixed in various patterns on a floor. Different finishes, such as honed and polished, can also be mixed. While these combinations can be very attractive, their maintenance and eventual restoration must also be considered. They can prove difficult to restore, and will be much more expensive to repair should that be required.

3. We are designing a master bath for our new house, and would like to use some type of a black stone on the floors, walls and vanity top. Should we use marble or granite?

First you need to determine how

often and how heavily the bath will be used. Black marbles are typically soft, easily scratched sensitive to acids; they are also easier to repair than granite, and less expensive. Granite, on the other hand, offers more resistance to acid, is much harder, and therefore resists scratching, but is more expensive and difficult to repair when damaged. If I were you, I would consult a stone expert before making your final decision.

4. We have a beautiful set of stone steps leading to our front door. They are polished and very slippery when it rains . Is there any treatment we can apply to make them less slippery? I am afraid someone will fall and break his neck.

There are several options for making the steps less slippery.

Option 1. The steps can be honed and the polish removed. The high-gloss shine will be gone, but the steps will be less slippery. Not all of each step needs to be honed; several narrow strips can be honed into the step.

Option 2. Slip tape can be purchased at your hardware or home center and attached to the steps. This tape can detract from the beauty of the stone, but will provide a safe, less slippery step.

Option 3. Several companies sell chemical treatments which once applied, will add slip resistance to the steps. Most of these treatments are acid-based, however, and will detract from the shine.

Option 4. There are many brands of coatings or waxes that can be applied to the steps for added slip resistance. These coatings will place a barrier on the stone which will have to be reapplied as it wears. Be sure the coating you choose is resistant to ultraviolet (UV) light, or the coating may turn yellow.

RESOURCE DIRECTORY

Marble & Tile

Resource Directory

The following directory is a list of manufacturers and distributors that supply stone and tile cleaners, equipment and restoration supplies. Many of the supplies and equipment mentioned in this book can be found at home centers, hardware stores, janitorial supply and rental yards. Check your local yellow page directory

Marble & Tile

Alpha Professional Tools
250 Braen Ave
Wyckoff, NJ 07481
201-447-4330
800-648-7229
Fax 201-447-1128

Aqua Mix Inc
9419 Ann St.
Sante Fe, CA 90670
310-946-6877
Fax 310-946-3462

Braxton Bragg
P.O. Box 5407
Knoxville, TN 37928
800-575-4401
Fax 800-915-5501

Brightstone Products
1636 240th St
Harbor City, CA 90710
800-899-7193
Fax 310-326-7971

The Butcher Company
120 Bartlett St
Marlboro, MA 01752-3013
508-481-5700
800-225-9475
Fax 508-485- 9998

Ceramseal
Boston Street
Middleton, MA 01949
800-7BOSTIK
800-356-4903
Fax 508-750-7212

Chad Schnidt & Associates
PO Box 558
Hull, MA 02045-0558
800-342-4533
Fax 617-925-1758

Eastern Marble Supply Co.
PO Box 392
Scotch Plains, NJ 07076
908-789-6400
Fax 908-789-95555

Fahmie Brothers
7458 La Jolla Blvd
La Jolla, CA 92037
619-456-0171
800-732-3353
619-456-3981

Gran Quartz
1804 C Montreal Circle
Tucker, GA 30085
404-621-9777
Fax 404-621-9771

HMK Stone Care
2585 Thrid Street
San Francisco, CA 94107
415-647-3086
Fax 415-647-3089

Hertron International Inc
306 Clanton Road
Charlotte, NC 28210
704-523-6722

Hild Floor Machine Co. Inc
5339 West Lake Street
Chicago, IL 60644
312-379-8558
Fax 312-379-1776

IPT inc
3535 W. Harmon Ave #F
Las Vegas, NV 89103
800-227-8831
Fax 702-895-9884

Marble & Tile

Leitch & Company
939 Harrison Street
San Francisco, CA 94107
415-421-8485
800-999-8485
Fax 415-247-0977

Marblelife Inc
6900 Haggerty Rd.
Suite 140
Canton, MI 48187
800-627-4569
Fax 313-455-6130

Miracle Sealants
12806 Schabarum Ave #A
Irwindale, CA 91706
800-350-1901
Fax 818-851-8932

MTA Corporation
PO Box 397
Manchaca, TX 78652
512-282-9300
Fax 512-282-1003

Multi-Seal Marble Care
1109 South Fremont Ave
Alhambra, CA 91803
818-282-5659
Fax 818-282-8516

Nilfisk of America
300 Technology Drive
Malvern, PA 19355
215-647-6420
Fax 215-647-6427

ProSoCo Inc
PO Box 171677
Kansas City, KS 66117
913-281-2700
Fax 913-281-4385

Regent Products Inc
4698 Honeygrove Rd #2
Virginia Beach, VA 23455
804-460-5011
800-624-8210
Fax 800-525-8481

Stone Care International
PO Box 703
Owings Mills, MD 21117-0703
800-839-1654
Fax 410-363-8288

Tile Pro
5300 West 127th Street
Alsip, IL 60658
312-785-2407
800-786-2433
Fax 312-389-3745

Vic International
PO Box 12610
Knoxville, TN 37912
615-947-2882
800-423-1634
Fax 800-242-1141

VMC Technical Assistance Corp
4879 Olson Drive
Dallas, TX 75227
214-381-8456
Fax 214-381-8333

Wood and Stone
10115 Residency Rd
Manassas, VA 22110
703-369-1236

Wholesale Stone Suppliers
PO Box 2365
Melbourne, FL 32902-2365
407-725-8885

GLOSSARY

A

ABRASIVE
A substance used for abrading, smoothing and polishing. Aluminum oxide, tin oxide and diamond discs are examples of abrasives.

ABSORPTION COEFFICIENT
A value given to determine the rate at which a stone will absorb a liquid.

ACID
A water soluble chemical with a pH less than 7. Some typical acids are hydrochloric, hydrofluoric, acetic, sulfuric, phosphoric, and oxalic.

ADA
An abbreviation for American Disability Act. See chapter 7 for more information.

AGGLOMERATE
A stone type composed of many pieces of stone held together with polyester or similar resins.

ALKALINE
A water soluble chemical with a pH greater than 7. Some typical alkaline chemicals are ammonia and sodium hydroxide.

ALUMINUM OXIDE
An abrasive made from alum. Used for grinding, honing and polishing stone.

ASTM
American Society for Testing Methods. See chapter 9.

ATTEAPULGITE
A type of clay powder used for poulticing stains from stone.

B

BISQUE
A special material applied to ceramic tile prior to firing for obtaining a glazed finish.

BLEEDING
The ability of a substance to migrate into a stone.

BLUESTONE
A dense, fine grained sandstone exhibiting a bluish gray color.

BOWING
A twisting out of shape caused by unequal stresses.

BRECCIATTED MARBLE
A marble containing angular fragments in a matrix of various minerals.

BULL NOSE
A convex rounding of a stone tile or slab; typically found on a counter or vanity top.

C

CALCIUM CARBONATE
The main mineral found in most marble.

CAULKING
An elastic adhesive or plastic polymer used to seal the joints of stone.

CHIP
A small piece of stone or the void left by same.

COF
Coefficient of Friction. Used in measuring slip resistance to express the ratio of forces required to move one surface over the other under given vertical force.

COQUINA
A sedimentary rock composed of loose shell fragments cemented together.

CORIAN
A plastic resin solid surface fabricated into countertops manufactured by Dupont.

COATING
A protective layer applied to the surface of stone. Waxes, floor finishes, acrylics, and polyurethanes are all coatings designed to provide protection, waterproofing, and luster.

CRACK
A running break or split in a rock.

CRYSTALLIZATION
A term used to describe a marble polishing process. It is sometimes called recrystallization or vitrification.

CULTURED MARBLE
A resin based material with a gel coat surface formulated to imitate marble.

D

DENSITY
The closeness and compactness of particles to each other within a material.

DIATOMACEOUS EARTH
A powder type used for poulticing stains from stone.

DOLOMITE
A non-metamorphic limestone that has been penetrated by magnesium carbonate and whose content consist of a least 40% dolomite.

E

EFFLORESCENCE
A deposit of salts found on the surface of stone carried by moisture from the setting materials. Efflorescence appears as a white powder residue.

EPOXY
A two part resin with superior adhesion properties used for joining and filling stone. Also used for grouting stone and tile.

ETCH
A rough, dull mark produced by acid eating away at a polished surface.

EXPANSION JOINT
A joint between stone designed to expand and contract to prevent cracking of the stone.

F

FABRICATE OR FABRICATION
To construct or shape stone into another shape. For example, to cut a stone slab into a table top is to fabricate a table top.

FELDSPAR
A mineral found in many granite rocks.

FILLING
The filling of holes and cracks in stone with cements, plastic resins, or shellac.

FINISH
The final surface applied to the face of the stone. A wax or coating may also be called a finish.

FLAMNED FINISH
A rough jagged finished found on granite surfaces achieved by passing a flame over the surface of the stone causing certain minerals to pop out.

FRACTURE
A fault or failure in a stone producing a break in the stone.

FULLERS EARTH
A clay used for poulticing stains from stone.

FURAN
A resin grout type used for industrial settings.

G

GLAZED
A hard shiny surface that is fired on the surface of many ceramic tiles.

GRANITE
A natural stone composed chiefly of 30% quartz and 60% feldspar.

GRINDING
The process of using abrasive discs to flatten a stone surface.

GRIT SIZE
Applies to the size and number of particles per square inch which are used to determine the abrasiveness of sandpapers, grinding stones etc. See chapter 6.

GROUT
The material used to fill in the joints between stone tile. Grout can be cement or plastic resins.

GUM FREEZE
A product used to spray on chewing gum, freezing it, so that it can be removed.

H

HAMMERED FINISH
A finish on stone surfaces applied by hammering the surface and imprinting the surface of the hammer face to the stone.

HEARTH
The floor part of a fireplace on which the fire is placed.

HONED
A term used to describe a smooth surface finish on stone or the process of achieving the finish.

HYDROPHOBIC
Water repelling; typically used to describe the properties of certain sealers used to impart water repellence to stone.

I

IGNEOUS
A geological term used to describe a type of stone formation. Granite is an example of an igneous stone.

IMPREGNATOR
Term used to describe a stone sealer that penetrates the stone surface and does not form a coating. An impregnator is often referred to as a penetrating sealer.

INORGANIC
Term used to describe substances composed of matter other than plant or animal. Minerals are inorganic.

J

JOINT
The space between stone panels or tiles. Referred to as a grout joint.

K

KAOLIN
A type of clay used for poulticing stains from stone.

L

LATEX
A type of liquid plastic added to grout providing increased strength and improved color retention.

LIMESTONE
A sedimentary rock composed of calcium carbonate.

LIPPAGE
A term used to describe uneven tiles; i.e., when one tile is higher or lower than the adjacent tile.

M

MARBLE
Any carbonate rock or limestone, granular to compact in texture, capable of taking a polish, may be referred to as marble

METHYL CELLULOSE
A powder type used for poulticing stains from stone. Refer to Chapter 1 for further information.

METHYL SILICATE
A type of silicone found in water based penetrating sealers.

MEXICAN TILE
A clay tile produced by baking in the sun.

MORTAR
A mixture of cement, lime, sand, and water. Mortar is often used as a setting bed for tile.

MOSAIC
The setting of small tiles in a pattern or a picture.

M. S. D. S.
An abbreviation for Material Safety Data Sheet

N

NATURAL CLEFT
A stone finish achieved by splitting the stone from the ground . No treatment is performed to it's surface leaving its natural finish.

NEUTRAL CLEANER
A water based cleaner with a pH of 7. Used for daily cleaning of tile and stone surfaces.

O

OLEOPHOBIC
Oil repelling; used to describe the properties of certain sealers used to impart oil repellency to stone.

ONYX
A medium to hard marble found in caves exhibiting a banded translucent layering.

ORGANIC
A substance derived from living organisms (plant or animal). Stains such as food are organic in nature.

OXALIC ACID
A strong crystalline acid used with marble polishing compounds to polish stone. The acid reacts with the calcium carbonate to form calcium oxalate which facilitates the polishing process.

OXIDIZE
To change a compound by increasing its electronegative change. When iron turns to rust, oxidation of the iron takes place.

P

PAVERS
A term interchangeable with tile. Paver generally refer to floor tile only.

PENETRATING SEALER
Term used to describe a sealer that penetrates the stone surface. A penetrating sealer does not form a coating. Also called an impregnator.

pH
The measure of the acidity and alkalinity of a solution. Numbers are assigned from 1 to 14. Seven indicates a neutral pH, neither acid nor alkaline. Numbers lower than 7 are acid, and numbers higher than 7 are alkaline.

POINTING
The filling of joints between stone tiles or panels. Also referred to as grouting.

PORCELAIN TILE
A tile produced by baking at extremely high temperatures.
POROSITY
The ability of a material to absorb gases and liquids.

POULTICE
A powder or cloth placed on a stain, which is designed to remove the stain by absorption.

Q

QUARRY
The name given to the place where stone is removed from the ground.

QUARRY TILE
An unglazed ceramic floor tile.

QUARTZ
A mineral found in granite and other igneous rocks.

QUARTZITE
Stone composed of quartz and/or sandstone.

R

RECRYSTALLIZATION
The term used to describe a marble polishing process. Also called crystallization or vitrification.

REFINISH
The process of returning the surface to its original finish.

RESTORATION
The process of putting back into a former state or condition.

RIVER ROCK
A flooring material composed of river rock and bonded together with urethane.

S

SALTILLO TILE
A name given to many Mexican tile. See Mexican tile.

SANDBLASTED
A finish achieved by spraying sand using a compressed air.

SANDED GROUT
Cement based grout containing sand particles.

SANDSTONE
A sedimentary rock consisting of quartz grains cemented together by silica or calcium carbonate.

SEPIOLITE
A powder type used for poulticing stains from stone. Refer to Chapter I for further information.

SETTING
The act of installing stone or tile.

SHELLSTONE
A sedimentary rock composed of shell and coral fragments. Also referred to as coquina.

SILANE
A silicone based chemical found in penetrating sealers for stone.

SILOXANE
A silicone based chemical found in penetrating sealers for stone.

SLAB
A slice of stone produced by sawing a large block of stone.

SLATE
A stone consisting of metamorphosed clay materials.

SLIP RESISTANCE
The term given to a surface that exhibits non slip characteristics.

SOAPSTONE
A very soft stone material composed of the mineral talc.

SOLVENT
A liquid substance capable of dissolving other substances. Although water is considered a solvent, the term solvent generally refers to water less substances.

SPALL
The condition that results when the stone surface splits or is broken off.

STONE SOAP
A stone cleaner containing soaps made from vegetable oils.

STRIPPING
The process of removing waxes or other coating. Generally performed with stripping chemicals and abrasive pads.

STUN
Term used to describe the white mark that results from the striking of a sharp object against certain stone.

SUBFLOOR
A rough floor to which the tile is placed onto. A subfloor can be a concrete slab or wood.

T

TALC
A soft mineral found in some stone. Talc is found in soapstone.

TERRAZZO
A poured floor composed of marble chips and Portland cement.

TEXTURED
A pattern produced on the surface of a stone.

THERMAL FINISH
A term interchangeable with flamned finish. See flamned finish.

TIN OXIDE
A fine abrasive powder used in stone polishing compounds.

TOOLED
A pattern produced on the surface of a stone with a tool or hammer.

TRAVERTINE
A marble formed in hot springs. Characterized by its numerous holes of varying size.

TREAD
The top walking surface of a step.

U

ULTRAVIOLET LIGHT(UV)
Invisible, short wave, light rays that react with certain minerals within the stone which may fade or alter a stones color.

V

VEIN
The colored markings in stone.

VITRIFICATION
A term used to describe a marble polishing process. It is sometimes called crystallization or recrystallization.

W

WALK OFF MAT
A rug or carpet like material placed at the entrances of a tile/stone floor. Designed to remove sand, grit and dirt from ones shoe.

WALL GROUT
A non-sanded cement grouting material.

WARPING
Twisting out of shape caused by unequal stresses or bowed because of relatively weak restraining forces on the opposite side of thermal exposure in thin marble tiles/veneers.

WATER SPOTTING
The spot produced on a stone surface from the sprinkling or dropping of water.

WAXING
The application of a wax finish to the surface of a stone or the filling of natural voids in stone.

WEATHERING
The wearing of a stone surface resulting from elements of the weather.

INDEX

Marble & Tile

Marble & Tile